ENGINEERED CARBON NANOTUBES AND NANOFIBROUS MATERIALS

Integrating Theory and Technique

AAP Research Notes on Nanoscience and Nanotechnology

ENGINEERED CARBON NANOTUBES AND NANOFIBROUS MATERIALS

Integrating Theory and Technique

Edited by

A. K. Haghi
Praveen K. M.
Sabu Thomas

APPLE
ACADEMIC
PRESS

Apple Academic Press Inc.
3333 Mistwell Crescent
Oakville, ON L6L 0A2 Canada

Apple Academic Press Inc.
9 Spinnaker Way
Waretown, NJ 08758 USA

© 2019 by Apple Academic Press, Inc.

First issued in paperback 2021

Exclusive worldwide distribution by CRC Press, a member of Taylor & Francis Group
No claim to original U.S. Government works

ISBN 13: 978-1-77463-395-3 (pbk)
ISBN 13: 978-1-77188-704-5 (hbk)

Library and Archives Canada Cataloguing in Publication

Engineered carbon nanotubes and nanofibrous materials : integrating theory and technique / edited by A.K. Haghi, Praveen K.M., Sabu Thomas.

(AAP research notes on nanoscience & nanotechnology book series)
Includes bibliographical references and index.
Issued in print and electronic formats.
ISBN 978-1-77188-704-5 (hardcover).--ISBN 978-1-351-04812-5 (PDF)

1. Carbon nanotubes--Industrial applications. 2. Nanofibers--Industrial applications.
I. Haghi, A. K., editor II. K. M., Praveen, editor III. Thomas, Sabu, editor
IV. Series: AAP research notes on nanoscience & nanotechnology book series

TA455.C3E54 2018   620.1'93   C2018-904594-9   C2018-904595-7

Library of Congress Cataloging-in-Publication Data

Names: Haghi, A. K., editor.

Title: Engineered carbon nanotubes and nanofibrous materials : integrating theory and technique / editors, A. K. Haghi, Praveen K. M., Sabu Thomas.

Description: Toronto ; New Jersey : Apple Academic Press, 2019. | Series: AAP research notes on nanoscience and nanotechnology | Includes bibliographical references and index.

Identifiers: LCCN 2018037189 (print) | LCCN 2018037824 (ebook) | ISBN 9781351048125 (ebook) | ISBN 9781771887045 (hardcover : alk. paper)

Subjects: LCSH: Nanocomposites (Materials) | Nanostructured materials. | Nanofibers. | Carbon nanotubes.

Classification: LCC TA418.9.N35 (ebook) | LCC TA418.9.N35 E537 2019 (print) | DDC 620.1/93--dc23

LC record available at https://lccn.loc.gov/2018037189

Apple Academic Press also publishes its books in a variety of electronic formats. Some content that appears in print may not be available in electronic format. For information about Apple Academic Press products, visit our website at **www.appleacademicpress.com** and the CRC Press website at **www.crcpress.com**

ABOUT THE AAP RESEARCH NOTES ON NANOSCIENCE & NANOTECHNOLOGY BOOK SERIES:

AAP Research Notes on Nanoscience & Nanotechnology reports on research development in the field of nanoscience and nanotechnology for academic institutes and industrial sectors interested in advanced research.

Editor-in-Chief: A. K. Haghi, PhD

Associate Member of University of Ottawa, Canada;
Member of Canadian Research and Development Center of Sciences and Cultures. Email: akhaghi@yahoo.com

BOOKS IN THE AAP RESEARCH NOTES ON NANOSCIENCE & NANOTECHNOLOGY BOOK SERIES:

- **Nanostructure, Nanosystems and Nanostructured Materials: Theory, Production, and Development**
  Editors: P. M. Sivakumar, PhD, Vladimir I. Kodolov, DSc, Gennady E. Zaikov, DSc, A. K. Haghi, PhD

- **Nanostructures, Nanomaterials, and Nanotechnologies to Nanoindustry**
  Editors: Vladimir I. Kodolov, DSc, Gennady E. Zaikov, DSc, and A. K. Haghi, PhD

- **Foundations of Nanotechnology: Volume 1: Pore Size in Carbon-Based Nano-Adsorbents**
  A. K. Haghi, PhD, Sabu Thomas, PhD, and Moein MehdiPour MirMahaleh

- **Foundations of Nanotechnology: Volume 2: Nanoelements Formation and Interaction**
  Sabu Thomas, PhD, Saeedeh Rafiei, Shima Maghsoodlou, and Arezo Afzali

- **Foundations of Nanotechnology: Volume 3: Mechanics of Carbon Nanotubes**
  Saeedeh Rafiei

- **Engineered Carbon Nanotubes and Nanofibrous Material: Integrating Theory and Technique**
  Editors: A. K. Haghi, PhD, Praveen K. M., and Sabu Thomas, PhD

- **Carbon Nanotubes and Nanoparticles: Current and Potential Applications**
  Editors: Alexander V. Vakhrushev, DSc, V. I. Kodolov, DSc, A. K. Haghi, PhD, and Suresh C. Ameta, PhD

- **Advances in Nanotechnology and the Environmental Sciences: Applications, Innovations, and Visions for the Future**
  Editors: Alexander V. Vakhrushev, DSc, Suresh C. Ameta, PhD, Heru Susanto, PhD, and A. K. Haghi, PhD

- **Chemical Nanoscience and Nanotechnology: New Materials and Modern Techniques**
  Editors: Francisco Torrens, PhD, A. K. Haghi, PhD, and Tanmoy Chakraborty, PhD

CONTENTS

ABOUT THE EDITORS

A. K. Haghi, PhD

Professor Emeritus of Engineering Sciences,
Editor-in-Chief, International Journal of Chemoinformatics and
Chemical Engineering and Polymers Research Journal;
Member, Canadian Research and Development Center of Sciences and
Cultures (CRDCSC), Montreal

A. K. Haghi, PhD, is the author and editor of 165 books as well as 1000 published papers in various journals and conference proceedings. Dr. Haghi has received several grants, consulted for a number of major corporations, and is a frequent speaker to national and international audiences. Since 1983, he served as professor at several universities. He is currently Editor-in-Chief of the *International Journal of Chemoinformatics and Chemical Engineering* and *Polymers Research Journal* and on the editorial boards of many international journals. He is also a member of the Canadian Research and Development Center of Sciences and Cultures (CRDCSC), Montreal, Quebec, Canada. He holds a BSc in urban and environmental engineering from the University of North Carolina (USA), an MSc in mechanical engineering from North Carolina A&T State University (USA), a DEA in applied mechanics, acoustics, and materials from the Université de Technologie de Compiègne (France), and a PhD in engineering sciences from Université de Franche-Comté (France).

Praveen K. M.

University of South Brittany (Université de Bretagne Sud)—Laboratory
IRDL PTR1, Research Center "Christiaan Huygens," Lorient, France

Praveen K. M. is an Assistant Professor of Mechanical Engineering at SAINTGITS College of Engineering, India. He is currently pursuing a PhD in Engineering Sciences at the University of South Brittany (Université de Bretagne Sud)—Laboratory IRDL PTR1, Research Center "Christiaan Huygens," in Lorient, France, in the area of coir-based polypropylene microcomposites and nanocomposites. He has published an international article in *Applied Surface Science* (Elsevier) and has also presented poster and conference papers at national and international conferences. He also has worked with the Jozef Stefan Institute, Ljubljana, Slovenia; Mahatma Gandhi University, India; and the Technical University in Liberec, Czech Republic. His current research interests include plasma modification of polymers, polymer composites for neutron shielding applications, and nanocellulose.

Sabu Thomas, PhD

Professor of Polymer Science and Technology and Honorary Director of the Centre for Nanoscience and Nanotechnology, Mahatma Gandhi University, Kottayam, India

Sabu Thomas, PhD, is a Professor of Polymer Science and Engineering at the School of Chemical Sciences and Director of the International and Inter University Centre for Nanoscience and Nanotechnology at Mahatma Gandhi University, Kottayam, Kerala, India. The research activities of Professor Thomas include surfaces and interfaces in multiphase polymer blend and composite systems; phase separation in polymer blends; compatibilization of immiscible polymer blends; thermoplastic elastomers; phase transitions in polymers; nanostructured polymer blends; macro-, micro-, and nanocomposites; polymer rheology; recycling; reactive extrusion; processing–morphology–property relationships in multiphase polymer systems; double networking of elastomers; natural fibers and green composites; rubber vulcanization; interpenetrating polymer networks; diffusion and transport; and polymer scaffolds for tissue engineering. He has supervised 68 PhD, 40 MPhil, and 45 Masters theses. He has three patents to his credit. He also received the coveted "Sukumar Maithy Award" for the best polymer researcher in the country for the year 2008. Very recently, Professor Thomas received the MRSI and CRSI medals for his excellent work. With over 600 publications to his credit and over 23,683 citations, with an h-index of 75, Dr. Thomas has been ranked fifth in India as one of the most productive scientists.

CONTRIBUTORS

Jiji Abraham
International and Inter University Centre for Nanoscience and Nanotechnology,
Mahatma Gandhi University, P.D. Hills, Kottayam 686560, Kerala, India

Sharif Ahmad
Materials Research Laboratory, Department of Chemistry, Jamia Millia Islamia, New Delhi 110025,
India. E-mail: Sharifahmad_jmi@yahoo.co.in

R. Ansari
Faculty of Engineering, University of Guilan, Rasht, Iran

Jaydeep Bhattacharya
Plant Nanobiotechnology Laboratory, School of Biotechnology, Jawaharlal Nehru University,
New Delhi 110067, India

M. Esmaeili
Faculty of Engineering, University of Guilan, Rasht, Iran

Soney C. George
Centre for Nanoscience and Nanotechnology, Amal Jyothi College of Engineering,
Kottayam 686560, Kerala, India

Anujit Ghosal
Plant Nanobiotechnology Laboratory, School of Biotechnology, Jawaharlal Nehru University,
New Delhi 110067, India. E-mail: anuj.ghosal@gmail.com

A. K. Haghi
Department of Textile Engineering, University of Guilan, Rasht, Iran. E-mail: akhaghi@yahoo.com

Sajid Iqbal
Materials Research Laboratory, Department of Chemistry, Jamia Millia Islamia,
New Delhi 110025, India

Nandakumar Kalarikkal
International and Inter University Centre for Nanoscience and Nanotechnology,
Mahatma Gandhi University, P.D. Hills, Kottayam 686560, Kerala, India
School of Pure and Applied Physics, Mahatma Gandhi University, Kottayam 686560, Kerala, India

Shohreh Kasaei
Department of Computer Engineering, Sharif University of Technology, Tehran, Iran

Bentolhoda Hadavi Moghadam
Department of Textile Engineering, University of Guilan, Rasht, Iran

Nahid Nishat
Inorganic Materials Research Laboratory, Department of Chemistry, Jamia Millia Islamia,
New Delhi, India

Sukanchan Palit
Department of Chemical Engineering, University of Petroleum and Energy Studies,
Post Office Bidholi via Prem Nagar, Dehradun 248007, Uttarakhand, India.
E-mail: sukanchan68@gmail.com; sukanchan92@gmail.com

Rangnath Ravi
Materials Research Laboratory, Department of Chemistry, Jamia Millia Islamia,
New Delhi 110025, India

Eram Sharmin
Department of Pharmaceutical Chemistry, College of Pharmacy, Umm Al-Qura University,
Makkah Al-Mukarramah, Saudi Arabia

Sabu Thomas
International and Inter University Centre for Nanoscience and Nanotechnology,
Mahatma Gandhi University, P.D. Hills, Kottayam 686560, Kerala, India

Fahmina Zafar
Inorganic Materials Research Laboratory, Department of Chemistry, Jamia Millia Islamia,
New Delhi, India. E-mail: fahmzafar@gmail.com

Hina Zafar
Department of Chemistry, Aligarh Muslim University, Aligarh, Uttar Pradesh, India

PREFACE

Carbon nanotubes, with their extraordinary engineering properties, have garnered much attention in the past 10 years.

With a broad range of potential applications, the scientific community is more motivated than ever to move beyond basic properties and explore the real issues associated with carbon nanotube-based applications.

Carbon nanotubes represent one of the most exciting research areas in modern science. These molecular-scale carbon tubes are the stiffest and strongest fibers known, with remarkable engineering properties and potential applications in a wide range of fields.

This volume offers an overview of the main research groups around the world that have been focusing their efforts on the exploitation of this intriguing material, with the purpose of inspiring young scientists to follow their pathway.

Presenting up-to-date literature citations that express the current state of the science, this book fully explores the development phase of carbon nanotube-based applications. It is a valuable resource for engineers, scientists, researchers, and professionals in a wide range of disciplines whose focus remains on the power and promise of carbon nanotubes.

The first eight chapters of this new book focuses solely on carbon nanotubes, covering some major advances made in recent years in this rapidly developing field.

This book:
- Examines the historical evolution of nanofabrication as well as the trends and underlying technology
- Gives an overview of the emerging technologies and cutting-edge research in nanofabrication
- Covers the basics of carbon nanotubes
- Includes a simplified description of different topics related to carbon nanotubes
- Covers properties, growth, processing techniques, and individual major application areas of carbon nanotubes

- Provides comprehensive coverage with up-to-date literature citations
- Considers the future scope of CNTs

LIST OF ABBREVIATIONS

AFD	average fiber diameter
AFM	atomic force microscopy
AOP	advanced oxidation processes
AR	Allura Red
ARS	angular-resolved scatter
BTEX	benzene, toluene, ethyl benzene, and p-xylene
CAD	computer-aided design
CGAL	Computational Geometry Algorithms Library
CNM	carbon nanomaterials
CNTs	carbon nanotubes
CP	calcium phosphate
CP	conducting polymer
CR	Congo red
CVD	chemical vapor deposition
DC	direct current
DMF	N-N, dimethylformamide
DMSO	dimethyl sulfoxide
DSC	differential scanning calorimetry
DVI	digital volumetric imaging
DWCNTs	double-wall CNTs
EBSD	3D electron back-scatter diffraction
EPA	The Environment Protection Agency
FE	finite element model
FIB-SEM	focused ion beam-scanning electron microscope
FULs	fullerenes
GN	graphene
GO	graphene oxide
GSH	glutathione
HA	humic acid
HA-MWCNTs	hydroxyapatite-multiwalled carbon nanotubes
ISS	interfacial shear strength
ITO	indium tin oxide
L-CNTs	lignin-grafted carbon nanotubes
LA	laser ablation
LA	lactic acid

LSCM	laser scanning confocal microscope
MD	molecular dynamics
MM	molecular mechanics
MWCNTs	multiwalled carbon nanotubes
ox-MWCNTs	oxidized multiwall carbon nanotubes
PAN	polyacrylonitrile
PANI	polyaniline
PCVD	plasma-enhanced CVD
PEDOT	poly(3,4-ethylenedioxythiophene)
PLA	polylactic acid
PMMA	poly(methyl methacrylate)
PNP	purine nucleoside phosphorylase
PPy	polypyrrole
PR	Ponceau 4R
PSS	polystyrene sulfonate
PTh	polythiophene
PU	polyurethane
PVA	polyvinyl alcohol
RGR	relative growth rate
ROS	reactive oxygen species
RVE	representative volume element
SBR	styrene-butadiene rubber
SEM	scanning electron microscopy
SIFT	scale-invariant feature transform
STL	stereo lithography
SWCNTs	single-wall CNTs
TIS	total integrated scatter
uricase-MWCNT	uricase-modified MWCNT
vdW	van der Waals
VOCs	volatile organic compounds
VTK	Visualization Toolkit
WHO	World Health Organization

CARBON NANOTUBES AND ITS APPLICATIONS IN DIVERSE AREAS OF SCIENCE AND ENGINEERING: A CRITICAL OVERVIEW

SUKANCHAN PALIT*

Department of Chemical Engineering, University of Petroleum and Energy Studies, Post Office Bidholi via Prem Nagar, Dehradun 248007, Uttarakhand, India

E-mail: sukanchan68@gmail.com; sukanchan92@gmail.com

ABSTRACT

Nanotechnology and nanoengineering are today witnessing one drastic challenge over another. The world of challenges and the vision in the field of application of carbon nanotubes are changing the face of human scientific endeavor. Human civilization and human scientific endeavor today stands in the midst of immense scientific vision and deep introspection. Mankind's immense scientific prowess, the vast and versatile area of nanomaterials, and the futuristic vision will all lead a long and visionary way in the true emancipation of nanotechnology today. In this chapter, the author delves deep into the unknown depths of the science of carbon nanotubes and gives a detailed analysis of the recent endeavor in the field of nanomaterials, carbon nanotubes, and the domain of nanotechnology as a whole. Scientific validation and technological vision are the needs of scientific research pursuit today. This chapter explores and investigates the vast scientific potential, the scientific success and the scientific profundity behind application of carbon nanotubes in diverse areas of engineering and science. Scientific vision behind carbon nanotubes applications are far-reaching in today's world of scientific rejuvenation. The extraordinary mechanical properties and unique electrical properties of carbon nano-tubes have resulted in immense research activities across the world since their groundbreaking discovery by Sumio Ijima of the NEC Corporation, Japan in the early 1990s. Science has been a colossus with a vast vision after that fascinating discovery. Although early research focused on growth and characterization, these interesting properties have vastly led to an increase in the number of investigations focused deeply on application development in the past 5 years. This chapter unravels the intricacies of scientific vision in carbon nanotubes applications in vast and versatile areas of scientific endeavor. This watershed text uncovers the scientific difficulties and deep barriers in the path toward deep realization in nano-technology today.

1.1 INTRODUCTION

Carbon nanotubes application and application of nanotechnology to human society are today entering into a new era of scientific vision and deep scientific comprehension. Mankind's immense scientific prowess, the vast technological vision, and the scientific validation will today go a long and visionary way in the true emancipation and true realization of nanoscience and nanotechnology to human society today. The utmost need of science and human society are validation and introspection. Carbon nanotubes (CNTs) are a type of nanomaterial which has immense scientific potential and poised toward vast scientific advancements. This chapter focuses on the immense scientific introspection and the vast and versatile scientific achievements in the field of CNTs, nanomaterials, and engineered nanomaterials. Science is a huge colossus today with a definite vision and a strong will power of its own. Nanotechnology is a branch of scientific research pursuit which is today crossing vast scientific frontiers. The intricacies of nanoscience, the vast scientific forbearance of nanotech-nology, and the vast scientific applications of CNTs are deeply explored and investigated in this chapter. Since the discovery of CNTs by Ijima in 1991, immense progress has been made toward many diverse applications, for example, materials and devices. Human society today stands in the midst of deep comprehension as nanotechnology surges ahead in the vast panorama of science and engineering. A wide window of innovation is slowly opening as the author in this chapter successfully attempts to reach to the readers the intricacies of nanotechnology and CNTs applications.

1.2 THE AIM AND OBJECTIVE OF THIS STUDY

Mankind and human civilization are today highly challenged and the vision of scientific endeavor in nanotechnology is groundbreaking. Today vast scientific frontiers are surpassed. The purpose and the vision of this chapter are to investigate in deep details the intricacies and barriers in technological applications of CNTs and nanomaterials to mankind. Scientific conscience and scientific acuity are the hallmarks of this well-researched chapter. Human scientific endeavor today stands in the midst of failures and upheavals as regards applications of nanotechnology. The challenges of science have a few answers today. The main upshot of this chapter is the vast investigation into CNTs application, the futuristic vision of nanotechnology, and the exploration into the world of engineered nanomaterials. Engineered nanomaterials are the present day scientific marvels. This technology needs to be deeply investigated. This area of scientific research pursuit needs to be reenvisioned and reemphasized with each step of scientific and academic rigor. Nanotechnology today has a fascinating medley of research questions and research avenues. The author in this chapter also targets the importance of applications of CNT applications in environmental engineering as well as chemical process engineering. Technological vision of nanoengineering today stands in the midst of scientific articulation and deep scientific acuity. The prime vision of this chapter is to explore and investigate the immense intricacies of research forays into CNTs, nanomaterials, and engineered nanomaterials. The challenge and vision of science and engineering are immensely groundbreaking today. This chapter is an eye-opener to the researchers and scientists with the sole vision of furtherance of science of nanotechnology.[1,2,13]

1.3 THE SCOPE OF THIS STUDY

Nanotechnology today is surpassing vast scientific frontiers. The vision of application of CNTs and other nanomaterials has no bounds. Technology and science needs to be reenvisioned and readjudicated as regards application of CNTs or other nanomaterials to human society. Environmental protection as regards application of nanomaterials or engineered nanomaterials needs to be reenvisaged and reemphasized with the passage of scientific history and time. This chapter poignantly explores the scientific acuity and the scientific vision behind the environmental health hazards behind the application of nanomaterials especially CNTs. The scope of the study is vastly visionary and far-reaching. Scientific benefits to human society as regards nanotechnology, nanomaterials, and engineered nanomaterials are the vast scopes of this well-researched study. Today, science and technology are huge colossus with a much groundbreaking vision of its own. Engineering science and its vast and versatile scientific and academic rigor are the cornerstones of this study. Sustainability as regards energy and environment are today linked to scientific forays in nanotechnology by an unsevered umbilical cord. In a similar manner, the domain of energy and environmental sustainability needs to be readdressed with the progress of scientific rigor. In today's world, progress in nanotechnology is directly linked with energy sustainability. Electrical power engineering and energy sustainability are two opposite sides of the visionary coin. Thus, the scope of this study is vast and varied. Energy sustainability, nanotechnology, CNTs, and diverse areas of engineering science are the future research trend and future recommendations of this well-researched chapter.[1,2,13]

1.4   WHAT DO YOU MEAN BY CNTs?

CNTs are allotropes of carbon with a nanostructure that can have a length to diameter ratio greater than 1,000,000. These cylindrical carbon molecules have novel properties that make them highly useful in many applications of nanotechnology. Human scientific advancements and the futuristic vision of nanotechnology are all today leading a long and visionary way in the true realization of diverse branches of science and technology. Formally derived from graphene sheet they exhibit unusual mechanical properties such as high toughness and high elastic moduli. Referring to their electronic structure, they effectively exhibit semiconducting as well as metallic behavior and thus encompass full range of properties important to furtherance of technology. Nanotubes are categorized as single-walled nanotubes and multiwalled nanotubes. Human scientific endeavor and deep cognitive ability of science are leading toward a newer era of scientific regeneration and scientific instinct. Tools have been developed to produce nanotubes in sizeable quantities, including arc discharge, laser ablation (LA), chemical vapor deposition (CVD), silane solution method, and flame synthesis method.[1-5]

CNTs and nanomaterials are the next generation smart and eco-efficient materials. Technology today has a few answers to the tremendous scientific potential of nanotechnology and CNTs. Science needs to be reenvisaged and reemphasized with the passage of scientific history, scientific validation, and visionary time frame.

1.5   SCIENTIFIC DOCTRINE BEHIND NANOTECHNOLOGY TODAY

Today nanotechnology is in the path of newer scientific regeneration and is replete with deep scientific vision and introspection. Scientific articulation, deep scientific acuity, and the futuristic vision of nanotechnology will all lead a long and visionary way in the true realization of science and engineering today. The doctrine of science is today far-reaching and surpassing vast and versatile scientific frontiers. Nanoscience and nanotechnology today are huge scientific colossus with a deep vision and willpower of its own. In this chapter, the author deeply comprehends the potential, the applications, and the success of nanomaterials and engineered nanomaterials with the sole vision of furtherance of science and engineering. Nanotechnology and nanomaterials are moving toward a newer era of scientific regeneration and deep scientific forbearance. The challenge and the vision of the application of nanotechnology in environmental engineering science and chemical process engineering are changing the face of human scientific endeavor. Royal Society Report, United Kingdom[13] discussed in deep details opportunities, uncertainties, and challenges in nanoscience and nanotechnologies. Today nanoscience and nanotechnologies are vastly seen as having huge scientific potential to benefit in many visionary areas of research and application.[13] These scientific forays are rapidly attracting increasing investments from Governments and businesses throughout the world. Human scientific vision and human scientific resurrection are in the path of a newer era of innovation and instinct. This report (2004) deeply discussed the current state of scientific knowledge about nanotechnology and specific applications of the new technologies.[13] Human scientific ingenuity needs to be reenvisioned and reenvisaged with the passage of human history and time.

1.6   SIGNIFICANT RESEARCH ENDEAVOR IN THE FIELD OF NANOTECHNOLOGY

Nanoscience and nanotechnology are today in the threshold of a newer era and a newer rejuvenation. The vision of nanoscience is today ever-growing and replete with forbearance and fortitude. Research questions and research pursuit in nanoscience and nanotechnology are witnessing immense challenges and drastic changes. Scientific research endeavor in the field of nanotechnology needs to be reenvisioned and reemphasized with the passage of scientific history and the visionary timeframe. Nanotechnology today is surpassing wide and vast visionary scientific boundaries. The challenge and the vision of nanotechnology need to be reenvisaged as science and engineering surges forward. Energy engineering, material science, and diverse areas of engineering science are undergoing immense scientific restructuring globally. Energy sustainability and environmental sustainability are moving toward a newer era of vision and emancipation. This chapter depicts poignantly the scientific success, the vast scientific research pursuit and the technological challenges behind the wide emancipation of nanotechnology, nanomaterials, and material science. Global nanotechnology vision today stands deeply in the midst of deep scientific introspection. Energy engineering and environmental engineering today are directly linked with nanotechnology by an unsevered umbilical cord. Nanomaterials for environmental protection needs to be reenvisaged and reemphasized as science and engineering surges forward toward a newer visionary era.

1.7   NANOVISION, NANOENGINEERING, AND HUMAN SOCIETY: A VISION FOR THE FUTURE

The progress of human civilization and academic rigor in the field of nanotechnology are today linked by an unsevered umbilical cord. Human society and human civilization today stands in the midst of disaster, introspection and scientific vision. Nanoscience and nanotechnology are revolutionary areas of science and engineering today. In a similar manner, nanomaterials and engineered nanomaterials are the most promising areas of scientific pursuit and scientific vision in today's world. CNTs are an avenue in scientific endeavor in nanomaterials. Scientific harmony, deep scientific cognizance, and scientific inspiration are the necessities of human society with human progress. Nanovision and nanoengineering are the imperatives of scientific research pursuit as human civilization moves forward. The vision for the future in today's scientific world is ever-growing and exceedingly groundbreaking. Nanotechnology is today the need of the society. Material science, electronics engineering, bioengineering, environmental engineering, and chemical process engineering are the avenues of scientific endeavor which has interfaces with nanotechnology. In each area of scientific avenues of human society, nanotechnology today has gained immense importance. Thus, the major thrust of human research pursuit should be toward greater emancipation of nanoscience and greater realization of nanotechnology. Nanovision is the major thrust area of human scientific endeavor globally. Mankind's immense scientific prowess, the vast technological forays, and the success of nanoscience will all lead a long and visionary way in the true realization and true emancipation of material science and nanomaterials today. CNTs encompasses scientific endeavor in nanomaterials. In this chapter, the author rigorously points out toward the scientific vision and the deep scientific profundity in technological applications of CNTs. Human scientific conscience and deep scientific instincts are in the midst of introspection today globally. Environmental engineering, chemical process engineering, and petroleum engineering science are ushering in a new era in the area of scientific emancipation. Global concerns for environmental disasters, loss of ecological biodiversity, and frequent environmental mishaps are changing the scientific panorama today. This chapter focuses on the success of application of nanotechnology in chemical process engineering, environmental engineering, and petroleum engineering science. Depletion of fossil fuel resources is an area of immense concern today. Scientific validation and deep scientific motivation are challenging the entire scientific scenario globally. Nanovision and nanoengineering are today ushering in a new age of scientific regeneration and vast scientific rejuvenation. The upshot of this paper is the immense scientific and technological emancipation behind CNTs application and nanotechnology contribution to human society. Today nanovision is deeply connected to scientific research pursuit in petroleum engineering science and chemical process technology. The challenge and the vision need to be reemphasized and reenvisaged with the progress of academic and scientific rigor in the field of nanomaterials.[1-4]

1.8   RECENT SCIENTIFIC RESEARCH PURSUIT IN CNTs

Recent scientific research pursuit in CNTs is vastly visionary and crossing scientific boundaries. Human civilization and human scientific endeavor are in the path of newer scientific regeneration. CNTs and its vast and varied applications are changing the face of scientific research pursuit today.

De Volder et al.[1] discussed with deep and cogent insight, present and future applications of CNTs. Technological vision and scientific motivation are at their level with the progress of scientific and academic rigor. Worldwide commercial interest in CNTs is reflected in the production capacity that presently exceeds several thousand tons per year. Recently, bulk CNT powders are incorporated in diverse commercial products ranging from rechargeable batteries, automotive parts, and sporting goods to boat hulls and water filters.[1] Advances in CNT synthesis, purification, and chemical modification are enabling integration of CNTs in thin-film electronics and large area coatings.[1] Scientific vision, scientific profundity, and vast scientific regeneration are the hallmark of CNT application to human society today.[1] CNTs are seamless cylinders of one or more layers of graphene (denoted single-wall, SWNT, or multiwall, MWNT), with open or closed ends. Perfect CNTs have all carbons bonded in a hexagonal lattice except at their ends; whereas, defects in mass-produced CNTs introduce pentagons, heptagons, and other imperfections in the sidewalls that generally degrade desired properties.[1] Diameters of SWNTs and MWNTs are typically 0.8–2 nm and 5–20 nm, respectively, although MWNT diameters can exceed 100 nm.[1] The genesis of widespread CNT research in the early 1990s was preceded in the 1980s by the first industrial synthesis of what are now known as MWNTs and documental evidence and research on hollow carbon nanofibers started in the 1950s.[1] The vast vision of science, the challenges of nanotechnology, and the futuristic vision of physics of nanotubes will all lead a long and visionary way in the true emancipation of science and engineering.[1] After the brilliant definition of nanotechnology propounded by Dr. Richard Feynman, an eminent American physicist, science and engineering ushered in a new era.[1] Technology and engineering science of nanotechnology after the visionary definition by Dr. Feynman opened up new windows of scientific innovation and deep scientific instinct. Most of the CNT production today is used in bulk composite materials and thin films, which rely on unorganized CNT architectures having limited properties. CVD is the dominant mode of high-volume CNT production and generally uses fluidized bed reactors that enable uniform gas diffusion and heat transfer to metal catalyst nanoparticles.[1] In this chapter, the author with deep foresight reviews the genesis and application of CNTs with the sole vision toward scientific emancipation.[1]

Ajayan et al.[2] deeply discussed with immense foresight applications of CNTs. Human scientific vision, civilization's scientific prowess, and the needs of science to human society will lead a long and visionary way in the true realization of nanoscience, nanotechnology, and engineering science. CNTs have attracted immense interest worldwide with the furtherance of science and engineering today.[2] The small dimensions, strength, and physical properties of these nanostructures make them a unique material with a whole range of promising and visionary applications. Material science applications of CNTs are revolutionizing the scientific landscape. In this chapter, the author rigorously describes the electronic and electrochemical applications of CNTs, nanotubes as mechanical reinforcements in high-performance composites, nanotube-based field emitters, and their use as nanoprobes in metrology and biological and chemical investigations, and as templates for the creation of other nanostructures.[2] Vision of science and scientific validation are of utmost need in the research pursuit in nanomaterials today. The author deeply comprehends the challenges that result from the point of view of manufacturing, process engineering, and cost considerations in CNTs applications. This watershed text deeply discusses the vast scientific potential, the acuity, and the scientific articulation behind CNTs application.

Baughman et al.[3] discussed with deep and cogent foresight, the scientific intricacies and the routes toward application of CNTs. The vision and the challenge of nanotechnology and nanomaterials are ever-growing and overcoming scientific hurdles today. Human scientific vision, mankind's immense scientific prowess, and the scientific needs of the human society will all lead a long and visionary way in the true realization of science and engineering science. Many potential applications have been proposed for CNTs, including conductive and high-strength composites; energy storage and energy conversion devices; sensors; field emission displays and radiation sources; hydrogen storage media; and nanometer-sized semiconductor devices; probes and interconnects.[3] Human civilization and deep scientific endeavor are today in the threshold of a newer scientific regeneration. There are two main types of CNTs that can have high structural perfection.[3] Single-walled nanotubes (SWNT) consist of a single graphite sheet seamlessly wrapped into a cylindrical tube. Multiwalled nanotubes comprise an array of such nanotubes that are concentrically nested like rings of a tree trunk. Technology of nanotubes is highly advanced today. Scientific enigma and scientific vision are the cornerstones of research pursuit in nanotechnology today. Despite structural similarity to a single sheet of graphite which is a semiconductor with zero band gap, SWNT may be either metallic or semiconducting, depending on the sheet direction about which the graphite sheet is rolled to form a nanotube cylinder.[3] This direction in the graphite sheet plane and the nanotube diameter are obtainable from a pair of integers (m, n) that denote the nanotube type. The electronic properties of perfect MWNT are rather similar to SWNT because the coupling between the cylinders is weak in MWNT. Technology and engineering science of nanotechnology are surpassing vast and versatile scientific frontiers. Human scientific advancements and human peregrinations are today in a state of introspection and vision. This chapter widely opens the immense scientific profundity and the scientific forays into the murky world of MWNT and SWNTs. The first realized major commercial application of MWNT is their use as electrically conducting components in polymer composites. The low loading levels and the nanofiber morphology of the MWNTs allow electrical conductivity to be achieved while avoiding and minimizing degradation of other performance aspects and the low melt flow viscosity needed for thin-walled molding applications.[3] The vision and the challenge of nanotubes application are slowly changing the vast scientific panorama of nanotechnology. This chapter vastly pronounces and veritably opens up a new chapter in the field of applied science and nanotechnology. The challenge of science needs to be envisioned.

Popov[4] discussed with deep and cogent insight, properties and applications of CNTs. CNTs are unique tubular structures of nanometer diameter and large length/diameter ratios. The scientific success, the vision, and the challenges of science today are opening up new windows of innovation and scientific instinct in decades to come.[4] The nanotubes may consist of one up to tens and hundreds of concentric shells of carbons with adjacent shells separation of approximately 0.34 nm.[4] The carbon networks of the shell are closely related to the honeycomb arrangement of the carbon atoms in the graphite sheet. The amazing mechanical and electronic properties of the nanotubes stem in their quasi one-dimensional structure and the graphite-like arrangement of the carbon atoms in the shell. This is a watershed text in the review papers on CNTs. Scientific research pursuit and deep scientific profundity are the veritable needs of human civilization's progress today.[4] This chapter opens up new avenues and new directions in the field of nanotechnology applications. This report is intended to summarize some of the major advancements in the field of carbon nanotube research both experimental and deeply theoretical in connection with the future applications of nanotubes and the vast domain of nanotechnology.[4] The challenge, the vision, and the future recommendations of nanotechnology are enlivened with deep conscience in this paper.

Odom et al.[5] deeply and lucidly discussed single-walled CNTs from fundamental concepts to new device innovations. The frontiers of science and technology are wide open today and ushering in a new era in nanoscience and nanotechnology. Single-walled CNTs are ideal systems for investigating the fundamental concepts in one-dimensional electronic systems and have the tremendous scientific potential to revolutionize many areas of nanoelectronics. Technological vision and scientific forays in CNTs are today changing the scientific panorama of human scientific endeavor and human progress.[5] The future of application of CNTs is wide and bright. Today, science and engineering of nanotechnology are in the path of newer scientific regeneration and deep scientific adjudication. The authors deeply clarify the scientific success, the futuristic vision, and the scientific profundity behind CNTs application.[5] In this chapter, scanning tunneling microscopy has been used to characterize the atomic structure and tunneling density of states of individual SWNTs. Detailed spectroscopic measurements showed one-dimensional singularities in the SWNT density of states for both metallic and semiconducting nanotubes.[5] Technology and engineering science of nanotubes are surpassing vast and versatile scientific frontiers. This chapter widely presents with immense scientific vision the new device innovations in the field of CNTs. The results of the spectroscopic measurements obtained were compared to and agree well with the theoretical predictions and tight-binding calculations. Human scientific endeavor is at its zenith as nanotechnology and material science uncovers the deep scientific intricacies and scientific hurdles. One-dimensional nanostructures, such as CNTs and nanowires are poised to become important building blocks for molecule-based electrical circuits.[5] These materials not only offer the potential to serve as interconnects between active molecular elements in a device but also have the ability to function as the device element itself.[5] In order to exploit the unique physical properties of these one-dimensional materials for technological applications, it is important to characterize in detail their intrinsic properties.[5] The intricate determination of the properties of individual nanostructures has immense scientific value. This chapter unravels the deep scientific hurdles and the vast scientific intricacies of CNTs characterization.[5] Scanning tunneling microscopy and spectroscopy offer the potential to probe whether structural changes in SWNT geometry produce distinct electronic properties, since these methods are capable of simultaneously resolving the atomic structure and electronic density states of a material. Thus, the utmost importance of this method and the vision of this study.[5] The STM studies indicate that one-dimensional SWNT exhibit richness in electronic behavior that may be exploited for molecular device applications. The vast scientific vision, the technological cognizance, and the deep scientific profundity in this research endeavor are changing the scientific landscape, ushering new future thoughts and newer futuristic vision.[5]

Seetharamappa et al.[6] depicted with deep scientific revelation the domain of CNTs and the scientific success of next generation of electronic materials. The application of nanotechnology and CNTs are today surpassing vast scientific frontiers. The recent interest in CNTs is a direct consequence of the synthesis of buckminsterfullerenes. This discovery paved the way for the new scientific regeneration in the field of nanotechnology.[6] The discovery of buckminsterfullerenes took place in 1985. Science and engineering witnessed tremendous challenges after that groundbreaking discovery. The discovery that carbon could form stable, ordered structures other than graphite and diamond urged many researchers around the world to search for other allotropes of carbon. Human scientific research pursuit and the futuristic vision of nanotechnology today will lead a long and visionary way toward true emancipation of applied science and engineering science today.[6] In 1990, there was another great discovery that C_{60} could be produced in a simple arc-evaporation apparatus readily available in most laboratories. Sumio Ijima discovered the fullerene-related CNTs in 1991 using a similar evaporator. Human scientific vision attained a remarkable level with the discovery of CNTs. Nanotubes are composed of sp^2 bonds, similar to those observed in graphite and they naturally align themselves in ropes held together by van der Waals forces.[6] CNTs are cylindrical carbon molecules with immensely novel properties (outstanding mechanical, electrical, thermal, and chemical properties; 100 times stronger than steel, best field emission emitters, can maintain current density of more than 10^{-9} A/cm^2, thermal conductivity comparable to diamond) which make them potentially useful in a wide variety of applications (e.g., optics, nanoelectronics, composite materials, conductive polymers, sensors, etc.). CNTs are of two types, namely, SWNTs and MWNTs.[6] SWNTs were discovered in 1993 and most of them have a diameter close to 1 nm, with a tube length that may be many thousand times larger and up to orders of centimeters. Scientific cognitive ability and the deep scientific understanding are the necessities of research pursuit and research validation today.[6] This chapter goes deeper into the murky depths of scientific grit and determination in the vast domain of CNTs. This is a watershed text which goes beyond scientific imagination and scientific inspiration with the sole vision of emancipation of science and engineering.[6]

Ministry of Environment and Food (Denmark)[7] deeply discussed with cogent insight CNTs and the associated risks to man and the environment. The aims of the report were: (1) a detailed introduction into the physico-chemical complexity of CNTs, (2) overview of the production volumes and capacities for producing the different types of commercial CNTs, (3) overview of the current and near-market CNT-based downstream and consumer products, (4) overview of reported occupational, consumer and environmental release and exposure to CNT, (5) human and eco-toxicity of nanomaterials, and (6) risk assessment of CNTs to human society.[7] Human society and human civilization today stands in the midst of deep scientific vision and deep comprehension. A review of the deleterious effects of nanomaterials to environment is needed as human civilization moves forward. This chapter clearly targets the immense success of applications of CNTs to human civilization. The vision of science in the field of CNTs application is deeply delineated in this chapter. This chapter discussed structure and physical characteristics of CNTs, synthesis, purification, coating and functionalization, CNTs categorization, characteristics of commercial CNT, human and environmental exposure of CNTs, and human toxicology of CNT.[7] Technology and engineering science of nano-technology are today crossing visionary scientific boundaries. This chapter gives a wider glimpse on the success, the potential, and the deep techno-logical profundity behind CNTs application to human society.[7]

McEuen[8] deeply discussed with lucid and cogent insight single-wall CNTs. The technological and scientific challenges in CNTs application are immense and groundbreaking today.[8] Solid state devices in which electron are confined to two-dimensional planes have provided some of the most exciting scientific and technological breakthroughs of the last 50 years. Solid state physics and solid state chemistry are two domains of scientific pursuit which needs to be reenvisioned with the passage of scientific history and visionary timeframe. From metal-oxide-silicon field effect transistors to high mobility gallium arsenide heterostructures, these devices have played an important role in the microelectronics revolution and are critical components in a wide array of products from computers to compact-disk players.[8] Microelectronics and solid state physics are two branches of scientific endeavor which are connected by an unsevered umbilical cord. The author discussed electronic structure of nanotubes, conductivity of nanotubes, nanotube transistors, nanotubes as one-dimen-sional metals, and new devices and geometries. Validation of science and deep scientific profundity are the necessities of science and engineering in today's world. This chapter gives a watershed glimpse of the success of science of CNTs with sole vision of furtherance of nanotechnology.[8]

Science and engineering of carbon nanotubes are entering into a new era of scientific regeneration and deep scientific foresight. The world of challenges in scientific research pursuit in nanotechnology needs to be reenvisioned and reenvisaged. The scientific needs of human society, the vast technological vision of nanotechnology, and the needs of diverse areas of engineering will all lead a long and visionary way in the true emanci-pation of nanoscience and nanotechnology. CNTs and its applications are gaining immense importance day by day. In this chapter, the author deeply comprehends the vast scientific success, the challenge of engineering and nanotechnology, and the futuristic vision of CNTs.

Balasubramanian et al.[9] deeply discussed with cogent foresight chemi-cally functionalized CNTs. Since their discovery, CNTs have attracted the attention of researchers, scientists, and technologists throughout the world. Scientific vision, scientific understanding, and deep scientific comprehension are the hallmark of research pursuit in CNTs today. This unbelievable interest stems from their outstanding structural, mechanical, and electronic properties. Structural marvels of CNTs are the hallmark of research pursuit in nanotechnology and material science today.[9] CNTs show strong application potential in electronics, scanning probe microscopy, chemical and biological sensing, reinforced concrete materials, and in many diverse areas. Human scientific endeavor and nanotechnology are today interlinked by an unsevered umbilical cord. Human vision and scientific research pursuit are at its helm today. Many research pursuits in CNTs are in technical realization. Technological prowess and scientific profundity are in the face of immense scientific understanding and deep vision.[9] CNTs are cylinder-shaped macromolecules with a radius as small as a few nanometers, which can be grown up to 20 cm in length. The technical marvels and masterpieces in the field of nanomaterials are opening up new windows of scientific innovation and scientific instinct in decades to come. The walls of these CNTs are made of a hexagonal lattice of carbon atoms analogous to the atomic planes of graphite. They are capped at their ends by one half of a fullerene-like molecule.[9]

Zhang et al.[10] discussed with immense lucidity principles and processes behind CNTs mass production. Technological challenges and deep vision are the hallmarks and cornerstones of scientific research and scientific motivation in CNTs today. Human society requires new materials for sustainability and CNTs are amongst the most important advanced materials. This review describes the state-of-the-art of CNT synthesis, with a deep focus on their mass production in industry. Science of nanotechnology and CNTs are changing the global scientific landscape.[10] At the nanoscale, the production of CNTs involves the self-assembly of carbon atoms into a one-dimensional tubular structure.[10] The authors in this paper describe how this synthesis can be achieved on the macroscopic scale in processes similar to the continuous ton-scale mass production of chemical products in the modern chemical industry.[10] Human vision and human scientific research pursuit are today in the path of scientific regeneration and rejuvenation.[10] This chapter deeply discusses and clarifies the scientific success, the scientific regeneration, and scientific profundity behind nanomaterials production. This overview discusses processing methods for high purity CNTs and the handling of heat and mass transfer problems.[10]

Joselevich et al.[11] discussed with immense lucidity carbon nanotube synthesis and organization. The synthesis, sorting, and organization of

CNTs are the hallmarks of deep scientific research endeavor in CNTs today. Mankind's deep scientific prowess, the futuristic vision, and the vast scientific profundity are the pallbearers toward a newer visionary era in nanotechnology.[11] This chapter reviews recent advances in these topics, addressing both the bulk production and processing of CNTs and their organization into ordered structures such as fillers and aligned arrays on surfaces.[11] The bulk synthetic methods are deeply reviewed with emphasis on the current advances toward mass production and selective synthesis. Human civilization's immense scientific prowess and regeneration will lead a long and visionary way in the true emancipation of nanotechnology and nanoscience today.[11] CNTs can have different individual structures, morphologies, and properties, as well as different collective arrangements and emerging properties, all of which are determined by the methods of preparation and chemical processing. In this chapter, the author deeply ponders and depicts with scientific ingenuity the success and vision of nanoscience and nanotechnology.[11]

Science and technology are today surpassing visionary boundaries. Technological foresight and adroitness are the cornerstones of human scientific research pursuit today. Today, nanotechnology and application of nanomaterials are changing the face of human scientific landscape. Human vision, the scientific far-sightedness and the futuristic vision will surely lead a long and visionary way in true emancipation of nanotechnology. The challenge of science today needs to be reemphasized and reenvisioned as regards application of CNTs or nanomaterials. In this chapter, the author rigorously points out toward the success and vision behind CNTs application.

1.9 APPLICATION OF NANOTECHNOLOGY AND CNTs IN ENVIRONMENTAL ENGINEERING

Scientific understanding and deep scientific discernment in the field of CNTs, nanomaterials, and nanotechnology are witnessing immense scientific rejuvenation today. CNTs can be veritably applied to diverse applications in engineering science including environmental engineering. Environmental protection and industrial pollution control today stands in the midst of vision and deep scientific introspection. Environmental engineering and disaster management are the cornerstones of research pursuit today. The state of environment of human planet today stands in the midst of ingenuity and crisis. Ren et al.[12] reviewed CNTs as adsorbents in environmental pollution management. CNTs have aroused a widespread global attention as a new type of adsorbent due to their outstanding ability for the removal of various inorganic and organic pollutants and radionuclides from large volumes of industrial wastewater.[12] This review summarizes the properties of CNTs and their properties related to the adsorption of various organic and inorganic pollutants from large volumes of aqueous solutions. Environmental protection and environmental engineering are moving from one scientific paradigm to another. CNTs have been termed as the materials of the 21st century.[12] Today technology of CNTs are widely visionary and immensely promising.[12] Environmental engineering and industrial wastewater treatment have diverse areas of scientific endeavor.[12] The application of adsorption is one such area. Human scientific endeavor and the vast academic rigor of unit operations in chemical engineering are changing the face of human civilization. CNTs have unique properties such as functional, mechanical, thermal, electrical, and optoelectronic properties.[12]

1.10 SCIENTIFIC SAGACITY AND DEEP VISION IN THE FIELD OF NANOTECHNOLOGY AND SUSTAINABILITY

The vision of science today in nanotechnology and sustainability are groundbreaking and deeply promising. Energy and environmental sustainability are the utmost needs of present day human civilization. Research trends today needs to be targeted toward successful application of sustainability toward human progress. Scientific and academic rigor in CNTs, nanomaterials, and engineered nanomaterials are changing the face of scientific genre today. The deep visionary definition of sustainability by Dr. Gro Harlem Brundtland, the former Prime Minister of Norway needs to be reemphasized and reenvisaged. Development of infrastructure in energy domain is of utmost need with the passage of scientific history and time. Today the vision of science is so ever-growing and the scientific challenges so arduous that research forays are of immense importance in the field of sustainable development. Deep concerns for environmental protection, the recurring issues of industrial wastewater treatment, and the concerns for ecological biodiversity have propelled the scientific domain to gear toward environmental sustainability. The success of environmental and energy sustainability are revamping the scientific mind-set and pushing the scientific domain toward newer innovations and newer forays in nanotechnology. Scientific sagacity today is in a state of immense distress. The author in this chapter poignantly depicts the vast scientific potential and the scientific acuity behind applications of nanomaterials and engineered nanomaterials with the sole vision toward greater furtherance of science and technology.

1.11   ENERGY SUSTAINABILITY AND THE VAST DOMAIN OF NANOTECHNOLOGY

Energy sustainability and nanotechnology are two opposite sides of the visionary scientific coin today. The scientific harmony and the scientific sagacity today stand in the midst of deep distress as well as introspection. Successful sustainability implies true realization of energy and environmental sustainability. Human scientific endeavor in nanotechnology is of utmost importance as regards human society's sustainable development. Nanotechnology is of immense necessity with the passage of scientific history and time. Application of nanomaterials and engineered nanomaterials in a similar manner needs to be reenvisioned and reenvisaged with the progress of scientific rigor. CNTs and other nanomaterials are of immense importance as regards sustainable development. The challenge and the vision of energy sustainability and electrical power engineering assumes in a similar vein immense importance. In this chapter, the author with deep and cogent insight depicts the need for nanotechnology in the furtherance of science as well as progress of human civilization. Energy engineering and the concerns for energy self-sufficiency are of immense importance in the path of scientific endeavor today.

1.12   APPLICATION OF NANOTECHNOLOGY IN DIVERSE AREAS OF ENGINEERING AND SCIENCE

Science and engineering are moving at a rapid pace in present day human civilization. Human scientific vision in the domain of nanotechnology in a similar manner is surpassing visionary scientific frontiers. Scientific and technological profundity is the utmost need of the hour as nanoscience and nanotechnology moves forward. Nanoscience and nanotechnology are today in the path of newer scientific rejuvenation and deep regeneration. CNTs and nanomaterials are witnessing a veritable scientific challenge. Nanotechnology today has an unsevered umbilical cord with chemical process engineering and the vast domain of environmental engineering science. Today the scientific world is faced with immense global issues such as loss of ecological biodiversity and frequent environmental disasters. Besides petroleum engineering and petroleum refining are witnessing a massive challenge that is the depletion of fossil fuel resources. In such a critical juncture of human history and time, nanoscience and nanotechnology assumes immense importance. Water science and technology and industrial wastewater treatment have deep scientific links with nanotechnology. The challenge of human scientific civilization and the human scientific endeavor are immense and groundbreaking today. Technology has a few answers to the success of environmental engineering tools and petroleum engineering techniques. This chapter repeatedly points toward the vast scientific success in the application of nanotechnology to diverse engineering avenues. Technology and engineering science are two huge colossi with a deep vision of its own. Industrial wastewater engineering, industrial pollution control, and the domain of drinking water treatment today stand in the midst of deep crisis and introspection. The challenges of these engineering techniques are depicted in details in this chapter.

1.13   VAST SCIENTIFIC DOCTRINE AND THE APPLICATION AREA OF NANOTECHNOLOGY

Vast scientific doctrine of nanotechnology is surpassing scientific and technological boundaries. The futuristic vision of nanotechnology applications, the immense scientific prowess of nanomaterials and engineered nanomaterials, and the scientific advancements will all lead a long and visionary way in the true emancipation of engineering and science today. Nanovision is the next generation hallmark of human progress. Technology and engineering science today are in the path of immense scientific regeneration and deep scientific vision. Scientific doctrine and scientific sagacity in nanotechnology applications today stands in the midst of introspection and profundity. This chapter deeply ponders the vast scientific success, the scientific revelation, and the scientific restructuring of the domain of nanotechnology. Nanotechnology and carbon nanomaterials applications are the challenges of human scientific progress. Royal Society Report,[13] envisioned and readdressed the environmental concerns and the health risks of nanomaterials application to human society. This report deeply discussed the deep scientific fortitude and scientific resurrection in the field of nanomaterials application. Today technology and engineering science of nanomaterials and nanotechnology are highly advanced and groundbreaking. Human scientific progress, the scientific and academic approaches behind nanotechnology applications will surely open up new windows of scientific innovation and scientific instinct in decades to come.

1.14 NANOMATERIALS FOR ENVIRONMENTAL PROTECTION AND THE VISION FOR THE FUTURE

Technology and engineering science are moving at a rapid pace today. CNTs in the similar manner are today ushering in a new era of scientific regeneration. Industrial pollution control, industrial wastewater treatment, and drinking water treatment are the utmost necessities of human vision today.

Hussain[14] deeply discussed with cogent insight carbon nanomaterials as adsorbents for environmental analysis. Nanotechnology can be defined as the science and engineering involved in the design, synthesis, characterization and application, organization of nanomaterials devices. Environmental analysis is the use of analytical chemistry techniques to measure and detect pollutants in the environment. Scientific vision and scientific profundity in the domain of environmental protection are crossing visionary boundaries. In this chapter, Hussain[14] depicts poignantly the scientific success, the scientific forbearance, and the scientific inspiration behind the applications of CNTs in environmental engineering science. Environmental analysis involves the determination of recalcitrant, natural, and harmful constituents in the environment. Carbon nanomaterials can be combined with other type of nanomaterials to form nanocomposites, thus incorporating different properties in a single new material.[14] They have unique electrical, optical, and mechanical properties which make them useful in the development of next generation of miniaturized, low power, and ubiquitous sensors. In addition to that, carbon nanomaterials can be used as nanoabsorbents for liquid as well as gas-phase adsorption of environmental pollutants because of their nanoscale adsorbent properties. The unit operation of chemical engineering such as adsorption is surpassing wide and vast visionary frontiers. Today adsorption of recalcitrant compounds in industrial wastewater is a major scientific success and crossing scientific frontiers. Among the carbon nanomaterials, CNTs, and fullerenes (FULs) receive the most attention because of their unique properties, which includes thermal conductivity, stability, tensile strength, and their ability to act as conductors and semiconductors.[14] CNTs which were first noticed by Ijima in 1991, have been applied for different purposes in the analytical science and analytical chemistry due to their mechanical, electrical, optical, and magnetic properties as well as their extremely large surface area. The synthesis of CNTs can be carried out by means of three main techniques: CVD, LA, and catalytic arc discharge.[14] Mankind's immense scientific prowess, the futuristic vision of science of nanotechnology, and the challenges of scientific endeavor will all lead a long and visionary way in the true realization of carbon nanomaterials applications today. The sorption sites on CNT are on the wall and the interstitial spaces between the tubes.[13] These sites are easily accessible to both rapid adsorption and rapid desorption. The impurities on CNTs reduce their availability because the sorbate has to diffuse through these impurities to reach the CNTs. FULs have attracted considerable attention since it was discovered in 1985. Technology and engineering science of FULs after the discovery moved forward with rapid pace into a new era of scientific rejuvenation. In this chapter, the author deeply discussed the physical, chemical, and biological properties of FULs and their diverse applications. It can be said that size, hydrophobicity, three-dimensionality, and electronic configurations make them an increasingly appealing subject in organic chemistry and applied chemistry. The author also discussed carbon nanomaterials (CNMs) for preconcentration of environmental pollutants and the vast techniques attached with it. Chromatographic applications of CNMs are also discussed in deep details.[14]

Ong et al.[15] discussed with lucid and cogent insight CNTs as next generation nanomaterials for clean water technologies. Water is an important resource for daily human society. The challenges of science and engineering of water technologies needs to be reenvisioned and reemphasized today. Water purification today stands in the midst of scientific inspiration and deep scientific resurrection. Although 70% of the earth's surface is covered with water, only 2% is fresh water.[15] Furthermore, 90% of the earth's fresh water supply is frozen in glaciers, which indicates limited availability of fresh water.[15] Research on the novel applications of CNTs in water treatment has shown a superior performance of CNTs in producing clean water. Technological vision and scientific validation are the veritable hallmarks of nanotechnology applications today. The authors in this chapter deeply discussed on the domain of adsorption of heavy metals. Heavy metals naturally exist in the ecosystem. However, the introduction of heavy metals to water by anthropogenic sources such as industrial and agricultural waste disposal has created a serious pollution problem that has vital global concerns. CNTs are recognized as a highly efficient adsorbent for removing heavy metals from water because of their large surface area and controlled pore size distribution.[15] Again surface-modified CNTs are more popular for adsorption studies than pristine CNTs because of their enhanced adsorption capability. The key area of adsorption of bacterial pathogens is another area of scientific endeavor in this chapter. The authors also dealt with life-cycle assessment of CNTs.[15]

Verma et al.[16] discussed with lucid insight engineered nanomaterials for purification and desalination of palatable water. Deep scientific understanding, scientific discernment, and scientific inspiration are the cornerstones of this chapter. Environmental protection and environmental engineering science today stands in the midst of vision and introspection. As per the United Nations Millennium Development Goals, access to safe drinking water, free from disease causing bacteria and viruses, is a basic human rights and is highly essential to the progress of human society. The authors in this chapter delved deep into the murky depths of environmental protection and tried to explore the possibilities of sustainable access to safe drinking water.[16] The applications of cutting edge tools of nanotechnology in environmental engineering are the other cornerstones of this study. Desalination is today a promising area of water purification technology to human society today. Nanotechnology could be used in water treatment for monitoring, desalinization, purification, and industrial wastewater treatment. The authors clearly targeted the application domain of ferritins, single-enzyme nanoparticles, CNTs, porous media and ceramics, nanofiltration, dendrimers, metal nanoparticles, and graphene in industrial and drinking water treatment.[16]

Science and engineering of nanomaterials or CNTs are widening the scope of scientific research in nanotechnology. The challenge and the vision of nanotechnology today need to be reenvisioned and reenvisaged.

Environmental protection globally is the need of the hour. In such a crucial juncture of scientific history and time, nanomaterials, and CNTs assumes vital importance.

1.15 FUTURE OF NANOTECHNOLOGY APPLICATIONS AND THE SCIENTIFIC VALIDATION

Nanomaterials and nanotechnology are the branches of science and engineering which needs to be envisioned with the passage of scientific history and time. Global scientific scenario in nanotechnology applications are ushering in a new era of scientific regeneration and the vast issue of scientific validation. Validation of science and engineering are of utmost necessity in the research pursuit in diverse areas of nanotechnology. Nanomaterials such as CNTs are reshaping the entire scientific domain of nanotechnology.[13] Today environmental risks of nanomaterials are of utmost concern as science and engineering moves forward. Future of nanotechnology applications needs to be targeted toward areas of research pursuit which is directly linked with human society.[13] Environmental protection and environmental engineering are one such area of scientific endeavor. Current and potential uses of nanotechnology are: (1) nanomaterials, (2) metrology, (3) electronics, optoelectronics and information and communication technologies, (4) bionanotechnology and nanomedicine, and (5) industrial applications.[13] Technology needs to be reenvisioned and reenvisaged as nanoscience moves forward. Research areas of stress in nanotechnology are:

- Development of suitable and practical methods for measurement of manufactured nanoparticles.[13]
- Investigation of methods of measuring the exposures of workers to manufactured nanoparticles.[13]
- Development of international standards and measures.[13]
- Establishment of global protocols for investigating the long-term fate of nanomaterials in environment.[13]
- In concurrence to research in groundwater remediation, development of an understanding of the transport and behavior of nanoparticles.[13]

- Epidemiological investigations of the interrelations of exposure and health outcomes in those industrial processes such as welding, carbon black, and titanium dioxide manufacture.
- Development of internationally agreed protocols and models for exploring the routes of exposure and toxicology to humans and nonhuman organisms of nanoparticles and nanotubes in indoor and outdoor environment.[13]
- A fundamental scientific understanding of interaction of nanoparticles with cells and their components with direct collaboration with environmental scientists, environmental engineers, and air pollution toxicologists.[13]
- Development of protocols for in vitro and in vivo toxicological studies of any new nanoparticles and nanotubes.[13]
- Adsorption through skin of different nanoparticles and the environmental risks behind it.[13]
- Determination of the risks of explosion of nanopowders.[13]

Human scientific vision and nanotechnology are today linked by an unsevered umbilical cord. Technology and engineering science are today moving at a rapid pace. The environmental issues of nanomaterial today is gaining immense importance as science and engineering moves forward.

1.16 FUTURE RECOMMENDATIONS AND FUTURE RESEARCH TRENDS

Future recommendations and future research trends in nanotechnology and CNTs should be more targeted toward applications in environmental engineering, chemical process engineering, and petroleum engineering. Science of nanotechnology today is a huge colossus with a definite vision and willpower of its own. Nanoscience and nanotechnology are today in a state of immense scientific regeneration and deep rejuvenation. Technology and engineering science needs to be reenvisioned with each step of scientific and academic rigor today. Human society and human mankind today stands in the midst of deep scientific vision and introspection. In this chapter, the author deeply targets the vast scientific success, the scientific potential, and the deep scientific comprehension behind CNTs applications. Future recommendations and future research trends should be directed toward greater futuristic vision and greater future applications of CNTs. Science today is a huge colossus with a vast and versatile vision of its own. This chapter targets the deep scientific truth and the scientific inspiration behind nanotubes applications to human scientific endeavor and human civilization. Research questions and research forays into nano-materials and the whole domain of nanotechnology should be targeted toward the development of human society taking into deep consideration the environmental engineering paradigm. Environmental protection, environmental engineering science, and industrial wastewater treatment are the utmost need of the hour. Nanotechnology and nanomaterials thus should be the cornerstone of visionary research endeavor. Energy sustainability is the other side of the visionary coin. Future recommendations and future research trends should follow and envision these scientific directions. Depletion of fossil fuel resources is another big hurdle to human progress. In such a crucial juncture, energy sustainability along with nanotechnology assumes immense importance. Technological challenges and validation of science are thus of immense relevance in today's world.

1.17   CONCLUSION AND FUTURE PERSPECTIVES

Science and engineering are moving forward with immense vision and scientific might. The challenge of CNTs application needs to be readdressed and reemphasized with the progress of scientific and academic rigor. In this chapter, the author rigorously points toward the imperatives of science and the technological profundity in the domain of CNTs application. Today human civilization and human scientific endeavor are in a state of immense scientific vision and scientific introspection. Technology and engineering science of nanotechnology are highly challenged and crossing vast and versatile boundaries. This chapter pointedly focuses on the diverse applications of CNTs in every branch of engineering and science. Environmental engineering science is one such area of deep scientific endeavor and vision. Nanovision is the next wide step of human civilization today. Environmental protection, industrial pollution control, and sustainability are the other areas of research pursuit in this chapter. Human civilization and human scientific endeavor today stands in the midst of deep scientific comprehension and vast scientific profundity. Future perspectives and future recommendations of this study should be targeted toward greater realization of sustainable development and greater emancipation of science and engineering. Present day human civilization needs to be revamped and restructured with human progress and progress of scientific rigor. This chapter poignantly ponders on the human success, the deep scientific forays into nanomaterials and engineered nanomaterials and futuristic vision of nanotechnology applications. Science and technology are today in the path of a new beginning and a newer innovative era. In such a critical juncture of human history and time, engineering science and technology needs to be reemphasized and reenvisaged with scientific and academic rigor of nanoscience and nanotechnology. The challenges, the barriers and the hurdles in the pursuit of nanotechnology are immense and ever-growing. This chapter is an eye-opener to the intricacies of nanotechnology and nanomaterials with a deep futuristic vision.

KEYWORDS

- **nanotechnology**
- **carbon**
- **nanotubes**
- **nanomaterials**
- **vision**

REFERENCES

1. De Volder, M. F. L.; Tawfick, S. H.; Baughman, R. H.; Hart, A. J. Carbon Nanotubes: Present and Future Commercial Applications. *Science* **2013**, *339*, 535–539.
2. Ajayan, P. M.; Zhou, O. Z. Application of Carbon Nanotubes. In *Carbon Nanotubes: Topics in Applied Physics*; Dresselhaus, M. S., Dresselhaus, G., Eds.; Springer-Verlag: Berlin, Heidelberg, Germany, 2001; Vol. 80, pp 391–425.
3. Baughman, R. H.; Zakhidov, A. A.; De Heer, W. A. Carbon Nanotubes: The Route Towards Applications. *Science* **2002**, *297*, 787–792.
4. Popov, V. N. Carbon Nanotubes: Properties and Application. *Mater. Sci. Eng. R* **2004**, *43*, 61–102.
5. Odom, T. W.; Huang, J.-L.; Lieber, C. M. Single-walled Carbon Nanotubes: From Fundamental Studies to New Device Concepts, *Ann. N Y Acad. Sci.* **2002**, *960*, 203–215.

6. Seetharamappa, J.; Yellapa, S.; D'Souza, F. Carbon Nanotubes: Next Generation of Electronic Materials. *Electrochem. Soc. Interface* **2006**, 23–61.

7. Environmental Protection Agency. *Carbon Nanotubes: Types, Products, Markets and Provisional Assessment of the Associated Risks to Man and the Environment*, Report of Ministry of Environment and Food of Denmark, 2015.

8. McEuen, P. L. Single Walled Carbon Nanotubes. In *Physics World*, 2000; pp 31–36.

9. Balasubramanian, K.; Burghard, M. Chemically Functionalized Carbon Nanotubes. *Small* **2005**, *2*, 180–192.

10. Zhang, Q.; Huang, J.-Q; Zhao, M.-Q.; Qian, W.-Z.; Wei, F. Carbon Nanotube Mass Production: Principles and Processes. *ChemSusChem* **2011**, *4*, 864–889.

11. Joselevich, E.; Dai, H.; Liu, J.; Hata, K.; Windle, A. H. Carbon Nanotube Synthesis and Organization. In *Carbon Nanotubes: Topics in Applied Physics*; Jorio, A., Dresselhaus, G., Dresselhaus, M. S., Eds.; Springer-Verlag: Berlin, Heidelberg, Germany, 2008; Vol. 111, pp 101–164.

12. Ren, X.; Chen, C.; Nagatsu, M.; Wang, X. Carbon Nanotubes as Adsorbents in Environmental Pollution Management: A Review. *Chem. Eng. J.* **2010**, *170*, 395–410.

13. Report of Royal Society. *Nanoscience and Nanotechnologies: Opportunities and Uncertainties*, Report of Royal Society, July 2004.

14. Hussain, C. M. Carbon Nanomaterials as Adsorbents for Environmental Analysis (Chapter 14). In *Nanomaterials for Environmental Protection*; Kharisov, B. I., Kharissova, O. V., Dias, H. V. R., Eds.; John Wiley and Sons: USA, 2014; pp 217–236.

15. Ong, Y. T.; Yee, K. F.; Yeang, Q. W.; Zein, S. H. S.; Tan, S. H. Carbon Nanotubes: Next Generation Nanomaterials for Clean Water Technologies (Chapter 8). In *Nanomaterials for Environmental Protection*; Kharisov, B. I., Kharissova, O. V., Dias, H. V. R., Eds.; John Wiley and Sons: USA, 2015; pp 127–137.

16. Verma, V. C.; Anand, S.; Gangwar, M.; Singh, S. K. Engineered Nanomaterials for Purification and Desalination of Palatable Water (Chapter 23). In *Nanomaterials for Environmental Protection*; Kharisov, B. I., Kharissova, O. V., Dias, H. V. R., Eds.; John Wiley and Sons: USA, 2014; pp 389–400.

CHAPTER 2

ENGINEERED NANOMATERIALS, NANOMATERIALS, AND CARBON NANOTUBES: A VISION FOR THE FUTURE

SUKANCHAN PALIT*

Department of Chemical Engineering, University of Petroleum and Energy Studies, Post Office Bidholi via Prem Nagar, Dehradun 248007, Uttarakhand, India

E-mail: sukanchan68@gmail.com; sukanchan92@gmail.com

ABSTRACT

The world of nanoscience and nanotechnology are surpassing one significant milestone over another. Scientific vision, deep scientific conscience, and the scientific fortitude will today lead a long and visionary way in the true emancipation of nanotechnology today. Human scientific and academic rigor in the field of nanotechnology is gearing today toward immense challenges. In this chapter, the author rigorously points out toward the scientific vision, the deep scientific cognizance, and the scientific profundity in the application areas of engineered nanomaterials, nanomaterials, and carbon nanotubes. Today nanotechnology is in the path of newer scientific rejuvenation and newer scientific restructuring. The author redefined the scientific vision behind application of nanomaterials and carbon nanomaterials with a clear objective toward furtherance of science and engineering. Today carbon nanotubes and other carbon nanomaterials have diverse applications in almost all the branch of science and engineering. Chemical process engineering, material science, petroleum engineering science, and environmental engineering are a few examples of this huge scientific vision of carbon nanomaterials and engineered nanomaterials applications. In this chapter, the author poignantly depicts and uncovers the scientific success, the scientific fortitude, and the vast scientific fabric of engineered nano-materials and its applications. Human scientific endeavor and the path of scientific profundity in the field of engineered nanomaterials are changing the vast scientific firmament today. This chapter is a deep scientific insight into the world of challenges and the world of scientific inspiration in the field of carbon nanotubes and engineered nanomaterials.

2.1 INTRODUCTION

Science and technology in today's world are witnessing immense hurdles, barriers, and deep introspection. Nanotechnology and nanoscience are changing the scientific fabric of present-day human civilization. Techno-logical marvels, the utmost needs of the human society, and the futuristic vision of science will all go a long and visionary way in the true eman-cipation of nanotechnology and material science today. Technology has a few answers to the advancements in engineered nanomaterials today. Engineered nanomaterials are relatively a new area of research pursuit today. In this chapter the author focuses on the deep farsightedness in the application of engineered nanomaterials and the broad domain of nanoma-terials in human society. Material science is a visionary area of scientific endeavor today. In this chapter, the author deeply comprehends the success of application of carbon nanotubes in the field of chemical process engi-neering, petroleum engineering, and environmental engineering science. The world of nanotechnology today is deeply linked with petroleum engi-neering science. Petroleum engineering today stands in the midst of deep scientific comprehension and vision. Depletion of fossil fuel resources is a major hurdle toward human progress today. Frequent environmental disas-ters and deepening environmental catastrophes in this century have urged the scientific domain to gear forward toward newer innovation and newer advances. The author rigorously points toward the scientific intricacies, the scientific hurdles, and the vision behind application of carbon nano-tubes and nanomaterials in the furtherance of science and engineering.

2.2 THE AIM AND THE OBJECTIVE OF THIS STUDY

Science and technology of nanomaterials and engineered nanomaterials are today in the midst of deep scientific vision and vast scientific introspection.

Carbon nanotubes need to be vastly envisioned and envisaged as regards its applications in diverse areas of science and engineering. Today, the science of nanotechnology is a huge colossus with a true vision and true forbearance of its own. Applications of carbon nanotubes in petroleum engineering, chemical process engineering, and environmental engineering are changing the face of scientific research pursuit today. The major aim and objective of this study is to target the diverse areas of applications in the field of carbon nanotubes, nanomaterials, and engineered nanomaterials. Today human civilization is a technology-driven human society. Scientific vision, deep scientific acuity, and the human needs are the hallmarks of this highly technology-driven society. The restructuring of science as regards nanotechnology and engineered nanomaterials applications to human society are the utmost needs of human civilization today. The primary objective of this well-researched chapter goes beyond scientific imagination as nanotechnology surges forward. The deep divination of science and the technological profundity will lead a long and visionary way in the true emancipation of nanotechnology and nanoengineering today. In such a crucial juncture of scientific history and time, nanomaterials and engineered nanomaterials research forays need to be focused with deep insight and scientific might. The author with deep scientific conscience deals with immense vision the intricacies of application of nanotechnology to human society and civilization.

2.3   WHAT DO YOU MEAN BY NANOMATERIALS AND ENGINEERED NANOMATERIALS?

Nanomaterials and engineered nanomaterials are the smart materials of today. Human scientific endeavor and scientific vision are today in a state of immense scientific regeneration. Nanotechnology today has applications in diverse areas of science and engineering. Technology today needs to be reenvisioned and restructured as regards petroleum engineering and environmental engineering science. The human civilization today stands in the crucial juncture of deep scientific introspection and vast challenges. Engineered nanomaterials, the vast domain of nanomaterials and its applications are changing the scientific firmament of nanotechnology, chemical process engineering, environmental engineering, and petroleum engineering. Engineered nanomaterials means intentionally created (in contrast with natural or incidentally formed) particle with one or more dimensions greater than 1 nm and less than 100 nm. Nanomaterials describe in principle materials of which a single unit is sized (in at least one-dimension) between 1 and 1000 nm but is generally 1–100 nm.

2.4 SCIENTIFIC DOCTRINE AND DEEP SCIENTIFIC VISION OF CARBON NANOTUBES

Carbon nanotubes and its applications today stand in the midst of deep scientific fortitude and vast scientific introspection. Human scientific progress today is in the path of newer scientific rejuvenation. Scientific doctrine, deep scientific cognizance, and the world of scientific challenges will all lead a long and visionary way in the true realization and the true emancipation of nanomaterials, engineered nanomaterials, and the vast domain of nanotechnology. The world of science and engineering are today facing immense scientific upheavals and visionary scientific barriers. Human engineering endeavor and human civilization's progress are today linked by an unsevered umbilical cord with the research forays in petroleum engineering and environmental engineering science. Nanotechnology vision and nanotechnology research forays are today challenging the scientific firmament of human civilization. Technological adroitness, scientific motivation, and deep scientific understanding are the hallmarks of scientific endeavor today. The world of science and technology today stands in the midst of hope and optimism as science and engineering of nanotechnology surges forward toward a newer visionary era. Carbon nanotubes and engineered nanomaterials are the necessities of human scientific progress in present-day human civilization. Scientific doctrine and scientific vision are challenging the firmament of nanotechnology today. In this chapter, the author rigorously points out toward the vast scientific vision, the scientific cognizance, and the deep scientific profundity in the application areas of nanotechnology in environmental engineering science and chemical process engineering.

2.5 THE CHALLENGE AND THE VISION OF APPLICATION OF NANOTECHNOLOGY

The challenge and the vision of application of nanotechnology in diverse areas of science and engineering are changing the face of human scientific endeavor. Scientific fortitude, scientific foresight and deep scientific discernment are challenging the fabric of engineering and technology today. Nanoscience and nanoengineering are the visionary and ground-breaking areas of science and technology today. Human scientific progress today stands in the midst of comprehension and vision. Nanotechnology and nanoengineering are today moving toward a newer scientific para-digm as human scientific progress enters into a new era. Nanoscience and nanotechnology are today linked to diverse areas of science and engi-neering by an unsevered umbilical cord. The vision, the targets, and the challenges of science and technology are today opening up new windows of innovation and new areas of scientific instinct in decades to come. In this well-researched chapter, the author deeply ponders on the vast scien-tific needs of nanotechnology applications to human scientific progress. The immense scientific prowess of human civilization, the vast scientific challenges and hurdles, and difficulties of scientific endeavor will lead a long and visionary way in the true emancipation of nanoscience and nanotechnology. Human scientific vision in the field of chemical process engineering, petroleum engineering, and environmental engineering are of utmost necessity in the path of scientific endeavor today. Nanovision is the need of science today. Scientific astuteness, scientific inspiration, and deep scientific profundity are the cornerstones of research pursuit in nanotechnology. The world of challenges in science today stands in the midst of deep vision and introspection. The author in this critical juncture of scientific history and time enumerates the deep scientific subtleties and technological steadfastness in the path toward emancipation of nanotech-nology applications.

2.6 SCIENTIFIC VISION, DEEP SCIENTIFIC FORBEARANCE, AND TECHNOLOGICAL VALIDATION OF ENGINEERED NANOMATERIALS

Engineered nanomaterials and nanotechnology are changing the true face of human scientific research pursuit. Technological prowess and innova-tion, scientific validation, and the futuristic vision of engineering science will all lead a long and visionary way in the true emancipation and true realization of nanoengineering and nanotechnology in today's world. Vali-dation of science in nanotechnology and carbon nanotubes applications are of immense importance in today's scientific path of regeneration and futuristic vision. Deep scientific forbearance and technological validation of engineered nanomaterials are of equal importance as science and engineering moves forward. Environmental engineering science, industrial wastewater treatment, and drinking water treatment are the utmost need of the hour. Vision and the challenge of engineered nanomaterials in water purification are of overarching importance in today's scientific world. Lack of pure drinking water and heavy metal groundwater contamination are veritably changing the vast scientific firmament. In the developing and developed nations throughout the world, there is a serious issue of lack of pure drinking water and heavy metal groundwater contamination. Science and technology of water treatment has a few answers to the burning issue of heavy metal and arsenic groundwater remediation. Environmental and energy sustainability is at a state of immense upheaval and scientific disaster. Today sustainable development has not reached every citizens of the human planet. Thus, global scientific order stands in the midst of scientific catastrophe and deep challenges. Industrial pollution control and water treatment are in the threshold of a newer vision and a newer era. In such a crucial juncture of human history and time, the need of nanotechnology and nanomaterials research is immense and groundbreaking. Engineered nanomaterials applications in environmental protection are also veritably the need of the hour. Human scientific endeavor in environmental pollution control is at a deep stake as human civilization surges forward. Science and its endeavor in nanomaterials are the challenging areas and the utmost need of the hour as civilization moves forward. This chapter truly depicts the success, the scientific potential, and the vast scientific imagination behind engineered nanomaterials and nanotechnology as a whole.

2.7   SIGNIFICANT SCIENTIFIC ENDEAVOR IN THE FIELD OF NANOMATERIALS

Scientific research pursuit in the field of nanomaterials and carbon nanotubes are changing the very basic and fundamental firmament of nanotechnology today. Applications of nanoscience and nanoengineering are changing the face of human civilization. Scientific endeavor needs to be reemphasized and reenvisaged with the passage of scientific history and time. Nanomaterials and engineered nanomaterials are today in the path of a new beginning. Scientific vision and deep scientific cognizance are the utmost need of the hour as nanotechnology moves forward toward a newer knowledge dimension. In this section, the author deeply discusses with cogent insight the significant endeavor in the field of nanomaterials especially application of nanomaterials for environmental protection. Environmental engineering science today stands in the midst of deep scientific vision and deep scientific comprehension. Environmental protection and application of nanomaterials are the hallmarks of scientific research pursuit today. In this section, the author deeply comprehends the scientific needs and profundity in the area of nanomaterials and environmental protection.

Hussain[1] discussed with deep and cogent insight the area of carbon nanomaterials as adsorbents for environmental analysis. Technological validation and deep scientific motivation are of utmost need in the pursuit of science today. Science today is veritably visionary and steadfast as regards true realization of nanotechnology applications. Nanotechnology can be defined as the science and engineering involved in the design, synthesis, characterization and application, and organization of devices of nanomaterials. In the past few decades, nanotechnology has grown by leaps and bounds and this multidisciplinary scientific paradigm is witnessing immense restructuring and revolutionary development.[1] Nanomaterials, with bodily structures less than 100 nm in one or more dimensions have attracted significant attention and vision from scientists in recent years, mainly due to their unique attractive, thermal, electronic, mechanical, and biological properties.[1] Their high surface to volume ratios, the possibility of surface functionalization, and favorable thermal properties provide the flexibility needed for a broad range of analytical as well as fundamental applications.[1] The challenges and the vision of nanotechnology applications are today replete with vision and forbearance. Environmental analysis is the use of analytical chemistry to measure and detect the presence of pollutants in the atmosphere. Carbon nanotubes are today extensively used as promising nanomaterials for future applications. They can be combined with other types of nanomaterials to form nanocomposites, thus instilling different properties in a single nanomaterial.[1] Technology and engineering science are today in the path of newer scientific rejuvenation.[1] The author deeply discussed carbon nanotubes, fullerenes, adsorption on carbon nanotubes, carbon nanotubes for preconcentration of environmental pollutants, and chromatographic applications of carbon nanotubes. The other facet of this chapter encompasses membrane applications of carbon nanotubes.[1] Membrane applications in environmental engineering science and chemical process engineering are the immediate imperatives of scientific research pursuit. Human civilization's immense scientific girth, scientific determination, and vast scientific steadfastness are the veritable hallmarks toward membrane science applications and the vast domain of industrial wastewater treatment.[1]

Badawy et al.[2] deeply delineate with cogent insight nanomaterials for the removal of volatile organic compounds (VOCs) from aqueous solutions. VOCs are naturally occurring or man-made hydrocarbons with high saturation vapor pressures (greater than 102 kPa) at room temperature conditions (25°C). Examples of volatile organic compounds include benzene, toluene, ethyl benzene, and p-xylene (BTEX), chlorobenzenes, and chlorinated alkenes.[2] Extensive use of VOCs in varied industrial processes results in the discharge of large compounds in the environment. Human scientific endeavor's immense scientific prowess and vision, the deep acuity of scientific vision, and the vast needs of global environmental protection are the pallbearers toward a greater visionary era in nanomaterials today.[2] Nanomaterials for environmental protection are moving today in the path toward newer scientific regeneration. In this chapter, the author vastly comprehends the application of nanomaterials for BTEX removal.[2] The authors deeply enumerated on the topic of nanocomposites for BTEX removal, nanomaterials for chlorobenzene removal, metal oxides for chlorobenzene removal, carbon nanotubes for chlorobenzene sorption, nanomaterials for chlorinated alkenes removal, nanomaterials for phenol removal, and the vast scientific impact of nanomaterials on VOC removal by other processes.[2] Scientific research pursuits in nanomaterials for environmental protection are today ushering in a new era in the field of nanotechnology. Mankind's deep scientific prowess stands in the midst of introspection and vision.[2] This chapter reviews the vast technological and scientific challenges in the path toward scientific emancipation of nanomaterials application and nanotechnology realization to human society. Today scientific endeavor is a marvel of human civilization. The authors in this chapter vastly comprehend the success, the vision, and the cognizance behind nanomaterials application.[2]

Kharisov et al.[3] discussed with deep foresight nanomaterials on the basis of chelating agents, metal complexes, and organometallics for environmental purposes. Chelating agents and their metal complexes are at the core of modern coordination chemistry and in their original nonnano forms were used for environmental protection, for instance, for increasing the solubility of heavy metals in soil and therefore in enhancing phytoextraction.[3] In the past 20 years, in accordance with general development trends in nanoscience and nanotechnology, a variety of nanomaterials and nanocomposites have been created. In this chapter, the authors with deep insight discussed available nanomaterials and nanocomposites containing metal complexes and a few organometallic compounds used for the improvement of the environment.[3] Technological validation and scientific vision are the hallmark of this chapter. Chelating agents are the organic compounds which are in the centerpiece of immense scientific introspection today. Today green chemistry and green technology are the utmost needs of scientific endeavor as environmental protection and environmental engineering science garners global importance.[3] This chapter gives a detailed idea of the application of chelating agents in environmental degradation. Science of green technology today stands in the midst of deep scientific advancement and wide scientific fortitude. In this chapter, the authors discussed elemental metals functionalized with chelating ligands. A few functionalized (ligand-capped) or supported metal nanoparticles (generally Au, Ag, Fe, and bimetallics formed from them), containing chelating ligands, have been effectively applied for remediation of toxic metals by their chelation, as sensors for metal cations.[3]

Yee et al.[4] delineated with deep details, water remediation using nanoparticle and nanocomposite membranes. Environmental engineering science and environmental protection today are in the path of newer scientific regeneration and newer scientific vision. The World Health Organization (WHO) has deeply set up stringent guidelines on the allowable concentration of chemicals in drinking water, without causing health issues. The challenges and the targets of environmental engineering needs to be reenvisioned and reenvisaged with the passage of scientific history and time. Green chemistry, nanoengineering, and the wide domain of environmental engineering will lead a long and visionary way toward a true realization and true emancipation of science and technology today. The world needs a strong visionary understanding of improved water quality, industrial wastewater treatment, and drinking water treatment. Utilization of nanomaterials in improving water quality and groundwater remediation today stands as a major scientific imperative and a visionary area of research pursuit.[4] Nanomaterials possess properties suitable for water remediation, such as large surface area, high sorbent capacities, and the ability to be functionalized to enhance affinity and selectivity. Scientific endeavor, scientific vision, and deep scientific cognizance are the necessities of technology and engineering science today.[4] The authors deeply discussed water remediation by nanoadsorbents via the adsorption techniques, water remediation by nanophotocatalysts via photocatalytic degradation and nanocomposite membranes in water remediation. The major thrust area is the widespread development of membranes with antifouling properties and membranes with antibiofouling properties. Validation of science and engineering are the hallmarks in today's scientific endeavor in membrane science and fouling. Synthesis of nanomaterials and application of engineered nanomaterials in environmental protection will deeply lead a long way in the true realization of nanotechnology applications today.[4] Human scientific research pursuits today stand in the face of major difficulties and hurdles. Nanocomposite membranes and application of nanomaterials are the veritable cornerstones of scientific endeavor. This chapter is well-researched endeavor and opens wide windows of scientific innovation and instinct in the field of membrane science and nanomaterials applications.[4]

Rickerby[5] enumerated with scientific insight nanostructured titanium dioxide for photocatalytic water treatment. Nanotechnology and nanoengineering are today in the path of newer scientific regeneration and deep scientific vision. Human civilization and human scientific endeavor are witnessing immense challenges and vast metamorphosis as regards technological validation. Photocatalytic water treatment encompasses nontraditional environmental engineering water treatment techniques. Heterogeneous photocatalysis has attracted immense scientific attention as a method for water treatment because it is extremely effective in both, degrading organic and inorganic chemical pollutants and destroying bacteria, viruses, and pathogens.[5] Additional applications include air purification, self-cleaning and sterilizing surfaces, and water photolysis. Anatase titanium dioxide is the most commonly employed photocatalytic material but its efficiency is limited by an absorption spectrum confined to the ultraviolet range and a veritably charge carrier recombination rate.[5]

The wide vision of science and engineering are today opening up new thoughts and new instincts in the field of advanced oxidation processes and integrated advanced oxidation processes.[5] Scientific profundity, scientific steadfastness, and vast scientific cognizance are the cornerstones of research pursuit today. Titanium dioxide exists in three naturally occurring phases: tetragonal rutile, tetragonal anatase, and orthorhombic brookite (rutile is the most stable phase except in the case of very small crystal sizes where anatase and brookite become more stable).[5] The application of surface science techniques has provided molecular-level insights into the basic mechanisms of photocatalytic phenomena. Human scientific rejuvenation and deep scientific research forays are the challenges of human civilization today. In photocatalytic phenomena, the key topics of interest are photon absorption, charge carrier transport, trapping and recombination, electron transfer dynamics, adsorption and the adsorbed state, photocatalytic reaction mechanisms, effects of inhibitors and promoters, and the phase and form of the catalyst. The authors enumerated photocatalytic reaction, synthesis techniques, strategies for increasing photoreactivity, controlled crystal growth, and vast and varied applications of photocatalysis in water treatment.[5] Industrial wastewater treatment, drinking water treatment, and the vast world of traditional and nontraditional environmental engineering tools are revolutionizing the very scientific firmament. This chapter opens up new paths of scientific rejuvenation and scientific profundity in decades to come.[5]

Ramanathan et al.[6] discussed with deep and cogent foresight green synthesis of nanomaterials using biological routes. Dr. Richard Feynman, the eminent physicist proclaimed with immense vision and scientific fortitude the new genesis of nanotechnology.[6] He laid the foundations of nanotechnology and opened up new future thoughts and new future directions in the field of applied physics and engineering science. To realize the true potential of nanotechnology, it was immensely essential to develop fabrication methods and fabrication protocols of nanomaterial synthesis and testing their potential applications. Human scientific research forays, the vast technological vision, and the validation of science will veritably lead a long and visionary way in the true emancipation of nanotechnology applications today. Bioremediation today stands in the midst of deep scientific challenges and the vision.[6] During the detailed study of nature's strategies involved in fabricating new and smart materials, one of the significant discoveries was the important role of microbial activity in transforming elements in the periodic table, which is a result of assimilatory, dissimilarity, or detoxification processes. In this chapter, the authors deeply and poignantly depicts upon the domain of bioremediation routes, the world of metal NPs, metal sulfide NPs, bioleaching, and the novel approaches toward biosynthesis. Biochemical pathways and the bioremediation routes are the cornerstones of human scientific endeavor today. Technological validation in bioremediation is changing the face of science and the scientific firmament. Today water science and technology and bioremediation are connected by an unsevered umbilical cord. In this chapter, several investigations on the biosynthesis of inorganic nanomaterials by bacteria, fungi, algae, yeasts, and plant extracts have been made.[6] The authors deeply and poignantly depicted nanoparticle biosynthesis and its mechanistic aspect. Human scientific prowess, the deep scientific forbearance, and the technological profundity are the pallbearers toward a newer redefinition and newer innovations of nanoengineering and nanotechnology today. This challenge is profoundly discussed and poignantly depicted with immense vision and forbearance in this chapter.[6]

The scientific marvels and deep scientific cognizance are today in the path of newer regeneration. Nanotechnology and nanoengineering are changing the face of human scientific research pursuit. The scientific imagination, the deep scientific cognizance, and the face of scientific firmament are today changing the vision of technology and engineering science. In this chapter, the authors deeply targets and pronounces the scientific success, the vast scientific potential, and the endless vision of nanomaterials application in diverse areas of science and engineering. The author just pinpoints on the technological profundity in the domain of environmental protection and the holistic world of environmental engineering science.

2.8   SIGNIFICANT AND VISIONARY SCIENTIFIC RESEARCH PURSUIT IN ENGINEERED NANOMATERIALS

Engineered nanomaterials are the smart materials of today. Technological validation and scientific motivation in the application of nanotechnology to human society are the cornerstones of human advancement today. Nanotechnology and engineered nanomaterials are opening up new windows of scientific innovation and scientific instinct in decades to come. Carbon nanotubes and fullerenes are the avenues of scientific research pursuit which needs to be envisioned and envisaged. Scientific steadfastness, scientific regeneration, and deep scientific comprehension are the hallmarks of research pursuit today. Chemical process engineering, environmental engineering, and petroleum engineering science are today in the path of newer scientific rejuvenation and newer scientific instinct. In this chapter, the author rigorously points toward the scientific success, the scientific profundity, and the deep scientific astuteness in the scientific forays in chemical process engineering and environmental engineering science.

Verma et al.[7] discussed with immense foresight the success and vision of engineering science of engineered nanomaterials. The authors targeted on the domain of engineered nanomaterials for purification and desalination of palatable water. Environmental engineering science and environmental protection today are the utmost needs of human civilization today. In this chapter, the authors tried to explore the possibilities of sustainable access to safe drinking water by adopting cutting-edge tools of nanotechnology and nanoengineering. Today, nanoenabled technologies can efficiently process surface and groundwater for drinking purpose.[7] The authors today discussed in deep details the domain of desalination, aquaporins, ferritins, single enzyme NPs, carbon nanotubes, porous media and ceramics, nanofiltration, dendrimers, the world of nanoremediation, and vast application area of nanoscale zero-valent iron.[7] These are the various areas of engineered nanomaterials. Scientific vision needs to be reenvisioned and reemphasized as regards application of nanotechnology in environmental protection. This chapter vastly opens up new concepts, new future thoughts, and the scientific inspiration and imagination behind environmental protection.[7]

Gerasimov[8] discussed with deep and cogent insight fuel cells with nanomaterials for ecologically pure transport. Among the key technologies for the transition from environmental unfriendly fossil fuel use to the hydrogen-based economy, fuel cells assumes immense importance and are important devices for direct conversion of chemical energy into electricity. The challenge and the vision of science are deeply reflected in this well-researched chapter.[8] A large work in the fuel cell technology is dedicated to proton exchange fuel cells as they are the most important fuel cells for vehicle applications. Human scientific vision, the technological challenges, and the vision to move forward will today lead a long and visionary way in the true emancipation of nanotechnology and electro-chemistry.[8] In this chapter, the author gives a brief review of the recent progress in the application of nanomaterials for the improvement of nanomaterials in the metamorphosis into commercial applications. Fuel cell applications are today the challenge and vision of nanotechnology and nanoengineering.[8] The authors deeply discussed with cogent insight the scientific success, the vast scientific potential, and the technological validation behind nanomaterials applications. The author touched upon fuel cell vehicles, nanomaterials as catalysts, nanomaterials as catalyst support, nanostructured membranes, and the vast world of nanomaterials for hydrogen storage.[8]

Science and engineering are today in the path of deep scientific regen-eration and a newer scientific vision. Human challenges, the utmost need of environmental engineering and the vast vision of nanotechnology today will veritably lead a long and visionary way in the true emancipation of engineering science. Man's immense scientific prowess, the vast scien-tific zeal and determination of mankind and the unmitigated difficulties of research forays are today the pallbearers toward a newer visionary era in nanotechnology and environmental engineering.

2.9   SCIENTIFIC RESEARCH ENDEAVOR AND APPLICATION OF CARBON NANOTUBES IN ENGINEERING SCIENCE

Carbon nanotubes are the next generation smart materials. The challenge and the vision of science and engineering today, need to be revamped as regards application of carbon nanotubes to environmental engineering science and chemical engineering. Today chemical process engineering and environmental engineering are in the path of newer rejuvenation as environmental protection concerns revamps the scientific firmament.

The extraordinary mechanical properties and unique electrical prop-erties of carbon nanotubes (CNTs) have stimulated extensive scientific research across the world since their discovery by Sumo Iijima of the NEC Corporation in the early 1990s.[9] Technology and engineering science today are in the path of newer scientific rejuvenation. Although early research focused on growth and characterization, these interesting and groundbreaking properties have led to an increasing number of application developments in the past decade. Meyyappan[9] in a deeply researched chapter discussed in details CNTs and its vast and varied applications. This is a watershed text in the field of CNTs. The chapter gives a wider view on every application of CNTs in engineering and science. The entire gamut of applications for CNTs is indeed wide ranging and far-reaching: nanoelectronics, quantum wire interconnects, field emission devices, composites, chemical sensors, biosensors, detectors, etc.[9] Technological innovations and the world of challenges are today ushering in a new era in science and engineering of nanotechnology. The chapter starts with the structure and properties of CNTs. In deeply understanding the properties, the modeling efforts had been far-reaching and groundbreaking and have uncovered many interesting properties, which were later verified by hard characterization experiments. Scientific vision, scientific judgment, and deep scientific understanding are the hallmarks of today's research forays in CNTs and nanotechnology. The author also dealt with two early techniques mainly that produced single-walled nanotubes, namely, arc synthesis and laser ablation.[9] Chemical vapor deposition (CVD) and other related techniques emerged later as a viable alternative for patterned growth, though CVD was widely used in early fiber development efforts in the 1970s and 1980s.[9] These chapters on growth are followed by a chapter devoted to a variety of imaging techniques and characterization. Important techniques such as Raman spectroscopy are also covered in this chapter.[9] The focus on applications starts with the use of single-walled and multiwalled CNTs in scanning probe microscopy.[9] Technological and scientific validation are at its pinnacle as nanotechnology surges forward toward a newer age of rejuvenation. The challenge and the vision of science of nanotechnology are today crossing visionary boundaries. This chapter depicts poignantly the scientific success, the deep scientific potential, and the vast scientific profundity behind CNTs applications. Scanning probe microscopy is a visionary scientific endeavor. In addition to imaging metallic, semiconducting, dielectric, and biological surfaces, these probes finds applications in semiconductor metrology such as profilometry and scanning probe lithography. Technological challenges are immense and vital as nanotechnology and CNTs applications surges ahead. One of the chapters in this book summarizes efforts to date on making CNT-based diodes and transistors and rigorously attempts to explain the behavior of these devices based on semiconductor device physics theories. This entire chapter broadly depicts the vast scientific success of CNTs applications to science and human society. The other facets of this chapter are the deep investigations into the world of chemical and physical sensors and the true realization of biosensors development.[9]

Han[10] deeply discussed with deep and cogent insight structures and properties of CNTs. The deep scientific vision behind application of CNTs and the surge of nanotechnology are depicted poignantly at every step of this scientific research pursuit. Since the discovery of CNTs by Iijima in 1991, immense progress has been made toward the applications in materials which are[10]:

- chemical and biological separations, purification, and catalysis,[10]
- energy storage such as hydrogen storage, fuel cells, and lithium battery,
- composites for coating, filling, and structural materials,[10]

and in devices which are:

- probes, sensors and actuators for molecular imaging, sensing, and manipulation,[10]
- transistors, memories, logic devices, and nanoelectronic devices,[10]
- field emission devices for X-ray instruments, flat panel displays, and other vacuum nanoelectronic applications.[10]

The immense challenges and the scientific vision behind structures and properties of CNTs are delineated in deep details in this chapter. The properties which are enumerated in details in this chapter are (1) electrical, (2) optical and optoelectronic, (3) mechanical and electromechanical, (4) magnetic and electromagnetic, (5) chemical and electrochemical, and (6) thermal and thermoelectric. Today the scientific challenges and the deep technological profundity are immense as science moves toward a newer generation of vision and rejuvenation. The author (Han)[10] deeply targets this scientific vision.[10]

Srivastava[11] deeply comprehended on the domain of computational nanotechnology of CNTs. The science and technology of nanoscale materials, devices and their applications in functionally graded materials, molecular electronics, nanocomputers, sensors, actuators, and molecular machines form the realm of the domain of nanoscience and nanotechnology.[11] Applied physics and applied mathematics today have an unsevered umbilical cord with the vision of nanotechnology applications today. At a few nanometer scale, the devices and systems sizes begin to reach the limit of 10 to 100s of atoms, where even new physical and chemical effects are vehemently observed and form the basis of a new generation of cutting-edge products based on ultimate miniaturization where extended atomic or molecular structures form the basic building blocks. The real progress of nanotechnology has also been spurred by the discovery of atomically precise nanoscale materials such as fullerenes in the mid-1980s and CNTs in the early 1990s.[11] Technology has highly advanced since then and opened new avenues of scientific insight in nanotechnology applications. Human scientific vision has advanced at a rapid pace since then. This chapter is an eye-opener toward the vast scientific potential in research forays in nanotechnology. The importance of computational nanotechnology-based simulations in advancing the frontiers for the next generation of nanostructured materials, devices, and applications is based on three reasons. First, the length and time scales of important nanoscale systems and phenomenon have shrunk to the level where they can be addressed with high-fidelity computer solutions and theoretical modeling.[11] Second, the accuracy in the atomistic and quantum-mechanical methods has increased to the extent that, in many cases, simulations have become predictive.[11] Third, the raw central processing unit power available for routine simulation and analysis continues to increase so that it is regularly feasible to introduce more and more "reality" in the simulation-based characterization and application design.[11] The scientific vision, the scientific profundity, and the deep scientific forbearance of CNTs applications to science and engineering are enumerated in this chapter with lucidity. The relevant research questions for modeling- and simulation-based investigations of CNTs are truly multiple length- and time-scale in nature.[11] At the atomistic level there are accurate semiclassical and quantum simulation methods that feed into the large-scale classical molecular dynamics simulations with 10s of millions of atoms, which can then be coupled to mesoscopic (few hundreds of a nanometer length) devices and systems. Human scientific endurance, scientific grit, and the futuristic vision will all lead a long way toward the true emancipation of nanotechnology science today.[11]

Moravsky et al.[12] discussed with deep and cogent insight growth of CNTs by arc discharge and laser ablation. CNTs produced from carbon vapor generated by arc discharge or laser ablation of graphite generally have fewer structural defects than those produced by other known techniques. This is due to the higher growth process temperature that ensures perfect annealing of defects in tubular graphene sheets.[12] Multiwalled nanotubes (MWNT) produced by these high-temperature methods are perfectly straight, in contrast to kinked tubes produced at low temperatures in metal-catalyzed CVD processes.[12] While the quality of low-temperature tubes can be improved by prolonged postsynthesis annealing at temperatures above 2000 K, the mechanical and electrical properties of arc produced metal-walled nanotubes remain far superior. Human scientific regeneration in the field of CNTs is thus entering into a new era.[12]

Meyyappan[13] deeply discussed with cogent insight growth in CVD and plasma-enhanced CVD (PCVD). Scientific vision and deep scientific forbearance are the cornerstones of this chapter. As the applications for CNTs range from nanoelectronics, sensors and field emitters to composites, reliable growth techniques capable of yielding high purity material in desirable quantities are immensely essential.[13]

Global concerns of environmental protection are in a state of disaster and introspection. Nanotechnology and environmental engineering science are two opposite sides of the visionary coin. Human scientific endeavor, the needs of human society and the global concerns for water purification have veritably urged the domain of science and technology to surge toward newer vision and newer scientific frontiers. This entire chapter deeply discussed the challenges and the targets of science and engineering in chemical process engineering and environmental engineering today. The author widely ensured the scientific success and the deep scientific vision behind nanotechnology applications in human society.[17,18]

2.10   ENERGY SUSTAINABILITY AND MATERIAL SCIENCE

Energy sustainability and environmental protection are the utmost need of human society today. Mankind's immense scientific prowess, the technological challenges, and the futuristic vision of scientific research pursuit will all lead a long and visionary way in the true realization of energy sustainability and material science today. According to the visionary words of Dr. Gro Harlem Brundtland, the former Prime Minister of Norway and the visionary proponent of "sustainability," sustainable development can really be achieved if human society deeply participates and has team building conscience toward the true realization of engineering and science.[14-16] Today nanotechnology is one of the branches of scientific endeavor which needs to be envisioned and needs to be linked with the science of sustainability. Self-sufficiency in energy and the need for water purification are the other imperatives of science which needs to be linked with the world of sustainability. Material science and engineering science are the two branches of scientific pursuit which also needs to be veritably linked with the avenues of sustainability science. The domain of material science is so ebullient and groundbreaking and surpassing vast and versatile scientific frontiers. Nanotechnology and material science are two opposite sides of the scientific coin today. Sustainable development, whether it is energy or environment are the necessities of human progress today. The burning and vexing issues of climate change, loss of ecological biodiversity, and the depletion of fossil fuel resources are changing the scientific firmament of vision and introspection. Energy engineering and computer science are revolutionizing the world of challenges in sustainable development. Engineered nanomaterials are the success of science and the success of human civilization.[14-16]

2.11 NANOTECHNOLOGY APPLICATIONS IN ENVIRONMENTAL ENGINEERING SCIENCE

Environmental engineering science and environmental protection today stands in the midst of deep scientific vision and scientific forbearance. Industrial wastewater treatment and drinking water treatment stands as an important and utmost need to human society. Today nanotechnology has veritably diverse applications in different areas of science and engineering. Nanofiltration is one such visionary directions of research endeavor. There are other significant directions in nanotechnology applications today. Nanotechnology and environmental engineering science are in the path of newer scientific reenvisioning and deep scientific regeneration. Membrane science, traditional and nontraditional environmental engineering tools are reshaping scientific endeavor in industrial wastewater treatment and water purification. Technology revamping and scientific validation are of utmost necessity as drinking water crisis destroys the vast environmental engineering scientific firmament. Climate change, loss of biological diversity, and frequent environmental engineering disasters are of great concern to the progress of human civilization and human scientific endeavor today.

2.12   NANOTECHNOLOGY ENDEAVOR AND APPLICATIONS IN CHEMICAL PROCESS ENGINEERING

Chemical process engineering and nanotechnology are today linked by an unsevered umbilical cord. Water process engineering and petroleum engineering science are in the path of immense scientific revamping today. Nanotechnology has today vast and varied applications in petroleum refining. Nanofiltration has visionary applications in water science and water process engineering. Water science and technology has deep applications in the world of nanomaterials and CNTs. In this chapter, the author rigorously points out toward the vast scientific success, the scientific potential, and the deep scientific regeneration in the field of chemical engineering. Unit operations of chemical engineering are the backbones of chemical engineering endeavor. Here, the author deeply comprehends the scientific vision behind unit operations with specific contribution from the world of nanotechnology.[14–18]

2.13   INDUSTRIAL WASTEWATER TREATMENT AND NANOTECHNOLOGY ENDEAVOR

Industrial wastewater treatment and drinking water treatment today stands in the midst of deep scientific discernment and deep comprehension. Today science and technology has advanced so much that it made rapid inroads into the world of nanotechnology and environmental engineering. The world today stands in the face of a major disaster as climate change and loss of ecological biodiversity devastates the scientific firmament. Human civilization and human scientific endeavor needs to be reenvisioned and readdressed with deep scientific belief in eradicating global water problems. Heavy metal and arsenic groundwater contamination are the curse and bane of human scientific endeavor today. Technology and engineering science have numerous answers to the water issues. Innovative technologies such as desalination, chemical oxidation techniques, traditional, and nontraditional environmental engineering tools today are the hallmarks of scientific forays today.[17,18] Arsenic groundwater contamination is world's largest environmental disaster causing immense problems to citizens of developing and poor nations on earth. Heavy metal groundwater contamination is a raging issue to the developed nations also. The crisis needs to be immediately mitigated as science surges forward. Here comes the scientific imperative of nanotechnology. The areas of nanotechnology, nanofiltration, and the world of membrane science need to be reemphasized in future scientific research pursuit. Nanofiltration is a burgeoning area of science and engineering today. Water purification and industrial wastewater treatment are the need of human civilization today. Technological profundity and scientific motivation are the hallmark toward greater emancipation of water process engineering today.[17,18]

2.14   SCIENTIFIC ENDEAVOR IN THE FIELD OF PETROLEUM ENGINEERING AND NANOTECHNOLOGY

Petroleum engineering and nanotechnology are today linked by an unsevered umbilical cord. Nanoscience and nanotechnology are revolutionizing the scientific landscape as science and engineering moves forward. Oil and gas industry are today reframing and revolutionizing the vast scientific fabric of nanotechnology. Human scientific regeneration today needs to be redefined as regards application of nanotechnology in petroleum engineering. Depletion of fossil fuel resources is of grave concern to human scientific progress today. Technology and engineering science has a few answers to the difficulties and scientific hurdles of petroleum exploration today. Thus comes the need of nanotechnology, nanomaterials, and CNTs. Mankind's immense scientific prowess, the urgent need of fossil fuel resources and the vast technological challenges will all lead a long and visionary way in the true realization and true emancipation of nanotechnology today. Petroleum engineering science and petroleum refining, stand today in the midst of a difficult crisis with the deep disaster of depletion of fossil fuel resources.[14-16] The challenges and the scientific steadfastness will today lead a long and visionary way in the realization of scientific truth and deep forbearance. In this chapter, the authors deeply discuss with alarm and vision the feasible solutions toward mitigation of petroleum engineering crisis. Civilization and human progress needs to be reenvisaged and reemphasized as regards energy sustainability and petroleum refining. Technology needs to be redrafted at this critical juncture of scientific history and time. The difficulties and hurdles of human scientific progress need to be readdressed as regards petroleum resources and the cause of renewable energy. Renewable energy is one of the plausible solutions toward the fossil fuel crisis today. Scientific regeneration and research forays in renewable resources and alternate energy sources are the cornerstones of research pursuit today.[17,18]

2.15   ENVIRONMENTAL AND ENERGY SUSTAINABILITY AND HUMAN PROGRESS

Human progress today veritably depends on sustainable development in energy and environmental domain. Infrastructural development in energy and environment are the visionary thoughts of the proponents of science of sustainability.[14–16] Human scientific progress stands in the midst of deep catastrophe and deep scientific introspection. The visionary statement and definition of sustainability by Dr. Gro Harlem Brundtland, the former Prime Minister of Norway needs to be reenvisioned and reenvisaged with the passage of scientific history and time. Environmental and energy sustainability are of immense necessity to the human progress today. Mankind's immense scientific prowess, the progress of science and technology, and the futuristic vision of engineering science will all lead a long and visionary way in the true emancipation of scientific vision in environmental engineering science. Science and technology are moving at a rapid pace in present-day human civilization. Infrastructural development as regards energy and environment are the immediate need of the hour. Human civilization and human scientific progress are today entirely dependent on progress in science and technology and sustainability. Water science and technology and industrial wastewater treatment are today one branch of science which needs to be reemphasized as regards true realization of sustainable development. In this chapter, the author ponders upon the application of nanotechnology and nanomaterials to human society. Here in this respect, environmental engineering and sustainability assumes immense importance.[17,18]

2.16   FUTURE RECOMMENDATIONS AND FUTURE FLOW OF THOUGHTS

Science, technology, and engineering science are today in a state of deep comprehension and scientific fortitude. Nanotechnology is today linked with every branch of science and engineering by an unsevered umbilical cord. Scientific vision is today replete with fortitude and resurrection. The future of nanotechnology is vast and versatile today. Human scientific endeavor in nanotechnology are today in the crossroads of vision and profundity. Nanotechnology, chemical process engineering, environmental engineering, and petroleum engineering are the branches of scientific endeavor which needs to be reenvisioned and reenvisaged with the passage of scientific history and time. Future recommendations in nanotechnology are vast and versatile. Technology in every decade in this century is widening and in the process of deep restructuring. Environmental engineering science and industrial pollution control stands in the midst of immense difficulties and barriers. Climate change, loss of ecological biodiversity and the immense problem of environmental disasters has veritably urged human civilization to move toward a new eon of scientific rejuvenation. Scientific vision in the field of nanomaterials and engineered nanomaterials needs to be revamped as science witnesses immense hurdles.[17,18]

2.17   CONCLUSION AND SCIENTIFIC PERSPECTIVES

Scientific research pursuit in the field of engineered nanomaterials are today in a state of immense scientific rejuvenation and vast scientific vision. Scientific perspectives in nanotechnology endeavor needs to surpass scientific frontiers and go beyond the boundaries toward chemical process engineering, environmental engineering, and petroleum engineering science. In this chapter, the author repeatedly pronounces on the tremendous scientific advancements in the field of nanomaterials, engineered nanomaterials, and CNTs. Human scientific endeavor will surely usher in a new era if there is a true emancipation of nanotechnology applications today. Today environmental engineering science and water purification stands in the midst of deep scientific vision and introspection. This challenge of provision of clean drinking water needs to be reenvisioned and reemphasized as the science of nanotechnology surges forward. Future perspectives in the world of science and engineering of nanotechnology are ushering in a new era in the field of basic and applied science. This chapter aims toward a newer visionary era in the field of nanomaterials and engineered nanomaterials applications to human society. Science of nanotechnology veritably needs to be envisioned as regards chemical engineering, environmental engineering, and the vast domain of petroleum engineering science. These areas of science and engineering are the cornerstones of global research pursuit today. Human challenges, the futuristic vision, and the success of scientific endeavor will surely lead a long and visionary way in the true emancipation of nanotechnology today.

KEYWORDS

- **nanomaterials**
- **environment**
- **carbon nanotubes**
- **vision**
- **science**

REFERENCES

1. Hussain, C. M. Carbon Nanomaterials as Adsorbents for Environmental Analysis (Chapter 14). In *Nanomaterials for Environmental Protection*; Kharisov, B. I., Kharissova, O. V., Dias, H. V. R., Eds.; John Wiley and Sons: USA, 2014; pp 217–236.
2. Badawy, A. E.; Salih, H. H. M. Nanomaterials for the Removal of Volatile Organic Compounds from Aqueous Solutions (Chapter 5). In *Nanomaterials for Environmental Protection*; Kharisov, B. I., Kharissova, O. V., Dias, H.V. R., Eds.; John Wiley and Sons: USA, 2014; pp 85–93.
3. Kharisov, B. I.; Kharissova, O. V.; Mendez, U. O. Nanomaterials on the Basis of Chelating Agents, Metal Complexes, and Organometallics for Environmental Purposes (Chapter 7). In *Nanomaterials for Environmental Protection*; Kharisov, B. I., Kharissova, O. V., Dias, H.V. R., Eds.; John Wiley and Sons: USA, 2014; pp 109–124.
4. Yee, K. F.; Yeang, Q. W.; Ong, Y. T.; Vadivelu, V. M.; Tan, S. H. Water Remediation Using Nanoparticle and Nanocomposite Membranes (Chapter 17). In *Nanomaterials for Environmental Protection*; Kharisov, B. I., Kharissova, O. V., Dias, H. V. R., Eds.; John Wiley and Sons: USA, 2014; pp 271–291.
5. Rickerby, D. G. Nanostructured Titanium Dioxide for Photocatalytic Water Treatment (Chapter 10). In *Nanomaterials for Environmental Protection*; Kharisov, B. I., Kharissova, O. V., Dias, H. V. R., Eds.; John Wiley and Sons: USA, 2014; pp 169–182.

6. Ramanathan, R.; Shukla. R.; Bhargava, S.K.; Bansal, V. Green Synthesis of Nanomaterials Using Biological Routes (Chapter 20). In *Nanomaterials for Environmental Protection*; Kharisov, B. I., Kharissova, O. V., Dias, H. V. R., Eds.; John Wiley and Sons: USA, 2014; pp 329–348.

7. Verma, V. C.; Anand, S.; Gangwar, M.; Singh, S. K. Engineered Nanomaterials for Purification and Desalination of Palatable Water (Chapter 23). In *Nanomaterials for Environmental Protection*; Kharisov, B. I., Kharissova, O. V., Dias, H. V. R., Eds.; John Wiley and Sons: USA, 2014; pp 389–400.

8. Gerasimov, G. Fuel Cells with Nanomaterials for Ecologically Pure Transport (Chapter 28). In *Nanomaterials for Environmental Protection*; Kharisov, B. I., Kharissova, O. V., Dias, H. V. R., Eds.; John Wiley and Sons: USA, 2014; pp 471–482.

9. Meyyappan, M. *Carbon Nanotubes: Science and Applications*; CRC Press: USA, 2005.

10. Han, J. Structures and Properties of Carbon Nanotubes (Chapter 1). In *Carbon Nanotubes: Science and Applications*; CRC Press: USA, 2005; pp 1–24.

11. Srivastava, D. Computational Nanotechnology of Carbon Nanotubes (Chapter 2). In *Carbon Nanotubes: Science and Applications*; CRC Press: USA, 2005; pp 25–63.

12. Moravsky, A. P.; Wexler, E. M.; Loutfy, R. O. Growth of Carbon Nanotubes by Arc Discharge and Laser Ablation (Chapter 3). In *Carbon Nanotubes: Science and Application*; CRC Press: USA, 2005; pp 65–97.

13. Meyyappan, M. Growth: CVD and PECVD (Chapter 4). In *Carbon Nanotubes: Science and Applications*; CRC Press: USA, 2005; pp 99–116.

14. Ong, Y. T.; Yee, K. F.; Yeang, Q. W.; Zein, S. H. S.; Tan, S. H. Carbon Nanotubes: Next Generation Nanomaterials for Clean Water Technologies (Chapter 8) In *Nanomaterials for Environmental Protection*; Kharisov, B. I., Kharissova, O. V., Dias, H. V. R., Eds.; John Wiley and Sons: USA, pp 127–137.

15. Report of Royal Society. *Nanoscience and Nanotechnologies: Opportunities and Uncertainties*, July 2004.

16. Environmental Protection Agency. *Carbon Nanotubes: Types, Products, Markets and Provisional Assessment of the Associated Risks to Man and the Environment*, Report of Ministry of Environment and Food of Denmark, 2015.

17. Cheryan, M. *Ultrafiltration and Microfiltration Handbook*; Technomic Publishing Company Inc.: USA, 1998.

18. Van der Bruggen, B.; Manttari, M.; Nystrom, M. Drawbacks of Applying Nanofiltration and How to Avoid Them: A Review. *Sep. Purif. Technol.* **2008**, *63*, 251–263.

CHAPTER 3

SURFACE CHARACTERISTICS OF IONIC LIQUID-MODIFIED MULTIWALLED CARBON NANOTUBE-BASED STYRENE-BUTADIENE RUBBER NANOCOMPOSITES: CONTACT ANGLE STUDIES

JIJI ABRAHAM[1], NANDAKUMAR KALARIKKAL[1,2], SONEY C. GEORGE[3], and SABU THOMAS[1,4*]

[1]*International and Inter University Centre for Nanoscience and Nanotechnology, Mahatma Gandhi University, P.D. Hills, Kottayam 686560, Kerala, India*

[2]*School of Pure and Applied Physics, Mahatma Gandhi University, Kottayam 686560, Kerala, India*

[3]*Centre for Nanoscience and Nanotechnology, Amal Jyothi College of Engineering, Kottayam 686560, Kerala, India*

[4]*School of Chemical Sciences, Mahatma Gandhi University, Kottayam 686560, Kerala, India*

Corresponding author. E-mail: sabupolymer@yahoo.com

ABSTRACT

Influence of noncovalent surface modification of multiwalled carbon nanotube (MWCNT) by ionic liquid on the surface characteristics and wetting behavior of nanocomposites were investigated using contact angle measurements with water and dimethyl sulfoxide. The surface energy, work of adhesion, polarity, spreading coefficient, interfacial energy, and interaction parameter were measured for all composites. This chapter is presented as (1) contact angle studies of styrene-butadiene rubber (SBR)/f-MWCNT nanocomposites—effect of MWCNT loading and (2) contact angle studies of SBR/f-MWCNT nanocomposites—effect of ionic liquid loading. The wetting property shows that contact angle of water droplets on a sample surface is increased from ~87° (SBR surface) to ~105° (SBR with 5 phr f-MWCNT). Enhancement in hydrophilicity of the SBR nanocomposite films has been observed at higher ionic liquid concentration is attributed to increase in surface energy due to the incorporation of polar groups on the films surface. Tunable surface characteristics are obtained by controlling the chemical structure and composition of the polymer surfaces.

3.1   INTRODUCTION

In polymer nanocomposites, nanoparticles offer enormous advantages due to their higher surface area to volume ratio, higher aspect ratio, improved adhesion between nanoparticles and polymer, and comparatively less amount of filler loading needed to attain equivalent advances in properties. In these materials, addition of a second phase to polymer matrix led to improvements in bulk and surface characteristics.[1] It is widely accepted that improvement of properties of polymer nanocomposites can be linked to structural alteration in polymer surface due to the interactions with filler. So an in-depth study of surface characteristics of polymer nanocomposites is warranted. Research on wetting phenomena of polymer nanocomposites is a hot research topic on current scenario since this has reflected in the final applications of materials. During wetting, the contact angle between a liquid and a solid is zero or close to zero that the liquid spreads over the solid easily. In nonwetting, the contact angle is greater than 90° so that the liquid tends to ball up and run off the surface easily. Nonwettable surface are preferred in electronic applications, whereas wettable surfaces are preferred in applications such as biocompatibility, printing, coating and oil recovery, membrane for reverse osmosis, detergency, ore flotation, and retention of pesticides on leaves.

Surface characteristics such as adhesion and wettability of polymer nanocomposites can be determined by measuring the contact angle. Technical information which are obtained from contact angle measurements include the surface free energy, interfacial free energy, polar and dispersion components of surface energy, hydrophobic–hydrophilic alterations, polar group orientations, and restructuring of the surface in long time contact with a liquid.[2] The wettability of the filler in rubber, the interfacial adhesion between filler and rubber as well as the reagglomeration of filler are mainly driven by surface energies of filler and rubber matrix.

To date, considerable effort has been focused on investigating the influence of surface energies of fillers and interfacial interactions. Recent developments in the area of wetting of polymer nanocomposites show the correlation between morphology and wetting characteristics. Tai et al. reported that surface-modified Si_3N_4 nanoparticles brings well physical and chemical properties to styrene-butadiene rubber (SBR) with enhanced hydrophobicity and decreased surface free energy. This was due to the effective control of agglomeration tendency of modified nanoparticles in SBR.[3] Comparison of surface roughness value obtained from AFM analysis and the water contact angle shows that hydrophilic nature of TiO_2-coated multiwalled carbon nanotube (MWCNT)-based polyethersulfone membrane as compared to neat polymer membrane was due to the low surface roughness value for former.[4] Wu et al. investigated the influence of the filler surface chemistry on the dispersion of fillers and interfacial adhesion in the SSBR matrix by contact angle measurements.[5] Hydrophilicity or hydrophobicity of the nanoparticles on the surface of the polymeric matrix with filler loading was well understood from the contact angle values. Jose et al. demonstrated a transition of the inorganic nanofiller-based cross-linked polyethylene composites from wetting region to nonwetting region with increase in filler content.[6]

Researchers have used various types of covalent or noncovalent modification in the fabrication of rubber nanocomposites to enhance the dispersion of fillers and strengthen the interfacial interaction between the filler and rubber matrix. In the present study in order to tune the surface characteristics, noncovalent functionalization of MWCNT by ionic liquid was adopted. Here, we investigate the wetting characteristics of SBR nanocomposites as a function of modified MWCNT loading and ionic liquid loading.

3.2 THEORY AND CALCULATIONS

The contact angle is measured as the tangent angle formed between a liquid drop and its supporting surface. The techniques for measuring contact angle have been reviewed in detail by Neumann and Good. According to Thomas Young, when a droplet of a liquid is formed on a flat solid surface, the balance on the three-phase interface is expressed by following equation

$$(\gamma_{lv})\cos\theta = (\gamma_{sv}) - (\gamma_{sl}) \qquad (3.1)$$

where γ_{lv}, γ_{sv}, and γ_{sl} represent the liquid–vapor, solid–vapor, and solid–liquid interfacial tensions, respectively, and θ is the contact angle between the liquid–air interface and the surface.[7]

In eq 3.1, γ_s and γ_{sl} are not agreeable to direct measurement. Plot of cos θ against the surface tension for a homologous series of liquids, γ_l, can be extrapolated to give a critical surface tension, γ_c, at which cos $\theta = 1$.[8] The γ_c has been taken as an approximate measure of the surface energy, γ_s, of the solid. But a major drawback of this consideration is that the precise value of γ_c be determined by the nature of liquids used. To overcome this difficulty, an appropriate method has been introduced by Fowkes considering solid dispersion forces using a geometric mean equation.[9] Modified Fowkes equation was developed by Owens and Wendt[10] and Kaelble[11] by assuming the polar attraction forces including surface forces. By using harmonic mean equation which combines both dispersion and polar forces, Wu[12] has got still a better agreement to obtain γ_s. Wu's approach has been verified by selecting two liquids of dissimilar polarity, that is, water and dimethyl sulfoxide (DMSO) for obtaining γ_s of polymers.

Wu's harmonic mean equations are

$$(1+\cos(\theta_w))r_w = 4[(r_w^d r_s^d / r_w^d + r_s^d) + (r_w^p r_s^p / r_w^p)] \qquad (3.2)$$

$$(1+\cos(\theta_d))rd = 4[(r_d^d r_s^d / r_d^d + r_s^d) + (r_d^p r_s^p / r_d^p)] \qquad (3.3)$$

where the superscripts d and p stand for contributions due to dispersion and polar forces, respectively. Data for water and DMSO were taken from the literature (water $\gamma_w = 72.8$ mJ/m², $\gamma_w^d = 21.8$ mJ/m²; $\gamma_w^p = 51.0$ mJ/m², DMSO $\gamma_d = 44$ mJ/m², $\gamma_d^d = 36$ mJ/m², $\gamma_d^p = 8$ mJ/m²). Dispersive and polar component of surface energy of the composites γ_s^d and γ_s^p for different compositions of SBR were determined by solving eqs 3.2 and 3.3.

Surface energy is the energy associated with the interface between two phases. If the solid–vapor interfacial energy is low, the tendency for spreading to eliminate the interface will be less. The total solid surface free energy as per Owens–Wendt theory is represented as

$$\gamma_s = \gamma^d_s + \gamma^p_s \tag{3.4}$$

The work of adhesion W_A is the work which must be done to separate two adjacent phases 1 and 2 of a liquid–liquid or liquid–solid phase boundary from one another. Conversely, it is the energy which is released in the process of wetting. The work of adhesion, W_A, was calculated using the following equation.

$$W_A = (1 + \cos\theta)\gamma_l \tag{3.5}$$

where γ_1 is the surface tension of the liquid used for the contact angle measurement.

The interfacial energy is defined as the energy necessary to form a unit area of the new interface in the system.

Dupre's equation was used to calculate the interfacial free energy γ_{s1}.

$$\gamma_{s1} = \gamma_s + \gamma_1 - W_A \tag{3.6}$$

The spreading coefficient "S_c" implies that a liquid will spontaneously wet and spread on the solid surface if the value is positive, whereas a negative value of "S_c" implies the lack of spontaneous wetting.

$$S_c = \gamma_s - \gamma_{sl} - \gamma_l \tag{3.7}$$

Better understanding of the degree of interaction between the test liquid and the polymer surface was obtained by measuring Girifalco–Goods interaction parameter, ϕ, between the polymer and the liquid was determined using the equation given below.

$$\varphi = \frac{r_l(1+\cos\theta)}{2(r_l r_s)^{1/2}} \tag{3.8}$$

A: CONTACT ANGLE STUDIES OF SBR/F-MWCNT NANOCOMPOSITES: EFFECT OF MWCNT LOADING

3.3   RESULTS AND DISCUSSION

3.3.1   CONTACT ANGLE

Figure 3.1 shows images of water drops in the surface of unfilled SBR and composites containing different loadings of f-MWCNT (MWCNT modified with ionic liquid in the ratio 1:1). It is worth noting that the contact angle for near SBR matrix was around 87° and it is relatively a hydrophobic polymer. It can be seen that contact angle values increased significantly with the addition of MWCNT up to 5 phr f-MWCNT; this indicates the less affinity of nanocomposites toward water and thereby increases the hydrophobic nature. This variation in contact angle is due to the lower surface energy and the hydrophobic nature of MWCNT.[13] Conversely, when larger amounts of f-MWCNTs were added to the SBR matrix, a switch from hydrophobic to less hydrophobic character was observed. This might be due to the presence of relatively more amount of ionic liquid for modification of higher amounts of MWCNT. All the parameters obtained from contact angle studies are summarized in Table 3.1 and Table 3.2.

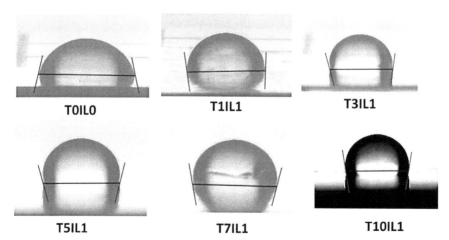

FIGURE 3.1   Contact angle measurements of SBR/f-MWCNT nanocomposites with water.

TABLE 3.1 Wetting Properties of SBR Nanocomposites of Composite–Water System.

Sample	Contact angle (°)	Work of adhesion (mJ/m²)	Interfacial energy (mJ/m²)	Spreading coefficient (mJ/m²)	Interaction parameter
T0IL0	87	76.39	20.32	−69.20	0.905
T1IL1	94	67.07	24.37	−78.52	0.895
T3IL1	101	58.06	28.65	−87.53	0.891
T5IL1	105	54.03	30.31	−91.07	0.880
T7IL1	102	57.26	28.83	−88.33	0.890
T10IL1	98	62.40	26.25	−83.19	0.893

TABLE 3.2 Wetting Properties of SBR Nanocomposites of Composite–DMSO System.

Sample	Contact angle (°)	Work of adhesion (mJ/m²)	Interfacial energy (mJ/m²)	Spreading coefficient (mJ/m²)	Interaction parameter
T0IL0	38	78.67	8.76	−14.80	0.958
T1IL1	43	76.17	8.97	−19.60	0.947
T3IL1	52	71.08	9.12	−27.99	0.933
T5IL1	59	66.66	9.34	−35.58	0.905
T7IL1	57	67.96	9.26	−33.58	0.912
T10IL1	53	70.47	9.08	−28.99	0.931

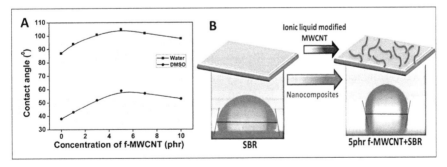

FIGURE 3.2 (A) Variation of contact angle of SBR/f-MWCNT nanocomposites with water and DMSO as a function of f-MWCNT loading. (B) Schematic diagram showing transition from wetting to nonwetting of SBR with the addition of f-MWCNT.

Variation of contact angle of SBR/f-MWCNT nanocomposites with water and DMSO as a function of f-MWCNT loading is depicted in Figure 3.2. Among the solvents used in the present study, water is polar in nature, whereas DMSO is slightly polar. Contact angle vary from solvent to solvent due to the polar and nonpolar nature of solvent and SBR. Contact angle value for water is more than that of DMSO due to the increased polarity of former as compared to latter and hence, nonwetting is observed with water. But for both solvents, contact angle values increases with increase in filler loading up to 5 phr MWCNT and decreases thereafter. SBR surface became more nonpolar with the addition of f-MWCNT and thus, hydrophobic character and contact angle is improved. Both uniform dispersion of f-MWCNT in rubber matrix and better filler–polymer inter-action facilitates the high packing density of macromolecules in the supra-molecular level. This creates kinetic and steric difficulties for diffusion of solvent molecules, which may result in high level of hydrophobicity.

3.3.2   SURFACE FREE ENERGY

Surface energy studies determine intermolecular interactions at the inter-faces of a solid surface with its environment. In rubber nanocomposites, physiochemical surface properties of rubber and filler have crucial role in various characteristics such as wettability of the filler by rubber, reag-glomeration of filler, and interfacial adhesion between filler and rubber. The thermodynamic contribution to interfacial interaction and the dispers-ibility can be calculated in terms of surface energies.[14] Surface energy of SBR nanocomposites with different filler loading using water and DMSO as solvents was calculated and is shown in Figure 3.3. Of the test liquids, DMSO was found to have the highest surface energy, whereas water has the lowest. Surface energy is the energy associated with the interface between two phases. Surface energy is found to be maximum in neat SBR (33.6 mJ/ m^2) and the total surface free energy shows a progressive decreasing trend with increasing MWCNT loading up to 5 phr f-MWCNT after that slight enhancement is observed. According to the concept of wetting process, hydrophobic nature is observed when the solid–vapor interfacial energy is low and then tendency for spreading to eliminate the interface will be less. In order to reduce the surface energy of the system, nanofillers should properly interact with polymer matrix. Composites with ionic liquid-modified MWCNT shows significantly decreased surface energy as compared to neat SBR which might be due to the decrease in free energy in mixing of two components at the interface.[15] The reduction in surface energy of the composite film with increase in f-MWCNT concentration is an indication of the tendency of the system to become less reactive with the surrounding when compared to the neat polymer. As the free energy of the system lowers, the polymer chains preferentially interact with the filler surface thereby decreasing the interaction with the surroundings.[16]

3.3.3 WORK OF ADHESION

The work of adhesion is the work required to separate the solid and liquid. The work of adhesion of composites was calculated using contact angle data because it depends on the contact angle and the surface tension of the liquid. Work of adhesion depends on the material of the substrate, the type of probe liquid and the temperature. Work of adhesion decreases as the surface free energy is decreased. The work of adhesion between solvent and polymer surface is monitored by the polar/nonpolar interactions and here as the f-MWCNT loading increases, nonpolar or hydrophobic

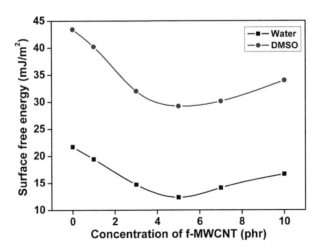

FIGURE 3.3   (See color insert.) Surface energy plots of SBR nanocomposites with water and DMSO as a function of filler concentration.

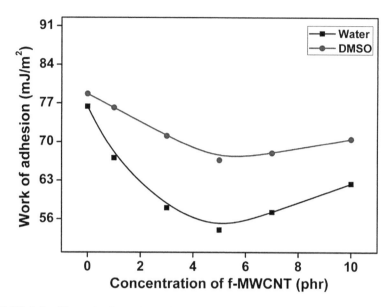

FIGURE 3.4 **(See color insert.)** Work of adhesion of SBR nanocomposites with water and DMSO as a function of filler concentration.

nature of composites decreases (Fig. 3.4). So the work of adhesion has decreased from 76.39 mJ/m^2 for SBR composite to 54.03 mJ/m^2 for T5IL1 composite, afterwards a slight increase is observed for polar solvent water. Work of adhesion can be correlated to the filler–polymer interaction. Effective filler–polymer interaction and the proper dispersion of fillers in the polymer matrix resulted in the reduction in work of adhesion values. The magnitude of difference between work of adhesion of SBR and that of composites are much higher for water than that of DMSO.

3.3.4 INTERFACIAL ENERGY

Figure 3.5 displays variation in interfacial energy of SBR nanocomposites with water and DMSO as a function of filler concentration. The interface between phases is a region of high energy relative to the bulk. The configuration of the surface adapts itself to minimize its excess energy to maintain the lowest total energy for the system and for the same, the dipoles orient themselves in such a way as to give minimum surface energy. Ions

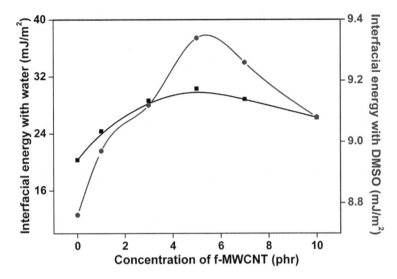

FIGURE 3.5 (**See color insert.**) Interfacial energy of SBR nanocomposites with water and DMSO as a function of filler concentration.

can fit on the surface with relatively low energy, only if they are highly polarizable ions, such that the electron shells can be distorted to minimize the energy increase produced by the surface configuration. Consequently, major fraction of the surface is covered with highly polarizable ions. Interfacial free energy is the excess free energy associated with an interface and this value is always less than that of sum of the separate surface energies of the two phases. In the present study, for both the solvents, interfacial energy increases with f-MWCNT loading up to 5 phr f-MWCNT-loaded nanocomposite. The increment in interfacial facial energy reveals that there is less chance for chemical attraction between the phases, solvent, and composites.

3.3.5 SPREADING COEFFICIENT

The spreading coefficient (S_c) indicates that a liquid will spontaneously wet and spread on the solid surface. If the spreading coefficient is positive, the liquid "spreads" on the solid; if the spreading coefficient is negative, the liquid "partially spreads" on the solid. If the contact angle is

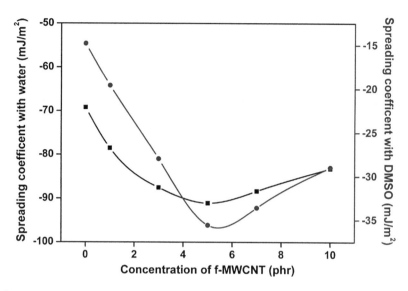

FIGURE 3.6 (See color insert.) Spreading coefficients of SBR nanocomposites with water and DMSO as a function of filler concentration.

less than 90°, the liquid is said to "wet" the solid; if the contact angle is greater than 90°, the liquid is said to "not wet" the solid. Positive values of spreading coefficient suggest spontaneous wetting, whereas negative values suggests no spontaneous wetting. Figure 3.6 displays the spreading coefficients of nanocomposites for water and DMSO. Spreading coefficients of both, water and DMSO decreases upon f-MWCNT addition. Wetting is quantitatively measured by polar–polar interaction across the interface between filler and rubber. On comparing the spreading coefficients values of two solvents, it is found that DMSO is a good wetting agent for the present system since it shows less negative values. In both the cases as the filler loading increases, Sc is shifted to higher negative values as compared to neat rubber which further confirms the hydrophobicity of prepared composites with filler loading.

3.3.6 INTERACTION PARAMETER

Degree of interaction between solvent and polymer surface is well understood by examining the Girifalco–Good's interaction parameter. Normally, the value of ϕ ranges from 0.5 to 1.15. The interaction parameter

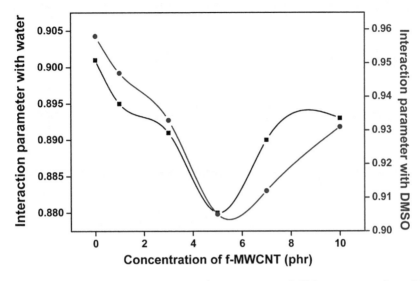

FIGURE 3.7 **(See color insert.)** Interaction parameters of SBR nanocomposites with water and DMSO as a function of filler concentration.

of nanocomposites due to two solvents is shown in Figure 3.7. Higher value of interaction parameter shows higher interaction between polymer surface and solvent. Highest value is obtained for neat SBR; it decreases upon filler addition and is in direct correlation with the filler content for both water and DMSO.

B: CONTACT ANGLE STUDIES OF SBR/F-MWCNT NANOCOMPOSITES: EFFECT OF IONIC LIQUID LOADING

TABLE 3.3 Wetting Properties of SBR Nanocomposites Containing with SME MWCNT with Different Amount of Ionic Liquid.

Sample	Contact angle (°)	Surface energy	Work of adhesion (mJ/m²)	Interfacial energy (mJ/m²)	Spreading coefficient (mJ/m²)	Interaction parameter
T3IL0	103	13.49	54.41	31.08	−90.12	0.873
T3IL1	101	14.71	58.06	24.37	−87.53	0.891
T3IL5	76	36.65	89.69	18.96	−55.90	0.896
T3IL10	67	43.45	100.32	15.13	−45.27	0.897

The water contact angles between ionic liquid-modified MWCNT and the rubber matrix, which can be seen in Table 3.3, clearly demonstrate a decrease in the contact angle with increasing the amount of modifying agent indicating the improvement in surface wettability. The SBR nanocomposites containing 3 phr unmodified MWCNT possessed the highest water contact angle of 103° while the water contact angles of the SBR/f-MWCNT nanocomposite membranes decreased significantly, reducing from 103 to 67 with the ratio between MWCNT and ionic liquid increasing from 1:0 to 1:10. In literature, some researchers suggested that good wettability of MWCNT by polymer matrix can be achieved by introducing active chemical groups on filler surface.[17,18] The nonwetting of MWCNT by deionized water can clearly indicate their hydrophobicity. The water contact angle of SBR composites with 3 phr unmodified MWCNT (T3IL0) was found to be around 103°, whereas modified MWCNT filled composites (T3IL5) showed a contact angle of around 76°. It can be speculated that ionic liquid modified MWCNT-networks endow quite hydrophilic behavior for modified MWCNT fillers compared to their inherent hydrophobic properties (Fig. 3.8). The addition of ionic liquid was expected to increase nitrogen content on membrane surfaces there by inducing decrease in contact angle. The increased hydrophilicity of f-MWCNT at higher amounts of ionic liquid is due to the surface coverage of hydrophilic moieties (nitrogen and bromine) in ionic liquid. The improved wettability of the nanocomposites with increase in ionic liquid concentration can be explained on the basis of decrease in surface roughness of the nanocomposites after surface modification (Fig. 3.8). Similar observation, that is, decrease in contact angle value with decrease in surface roughness value was made by several researchers.[19,20]

Surface free energy values of nanocomposites containing 3 phr MWCNT modified with different amounts of ionic liquid is displayed in Table 3.3. The surface free energy shows an obvious increase with increase in the amount of ionic liquid due to improved functional groups on the surface of MWCNT supplied by ionic liquid. The increase in interfacial wettability in systems with f-MWCNT can be attributed to a change of charge or polarity of the composite surface produced by the nanotubes.[21] Work of adhesion of nanocomposites showed a considerable enhancement of the hydrophilic nature of the MWCNT surface through the corresponding decrease in the water contact angle. The increase in work of adhesion with ionic liquid concentration was essentially due to the enhancement in its

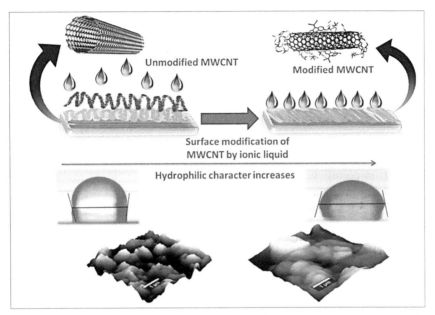

FIGURE 3.8 Schematic diagram showing transition from no-wetting to wetting characteristics with increase in the ionic liquid loading.

polar component because of the masking of hydrophobic part of MWCNT by the thin ionic liquid coating. Nanocomposites exhibit decreasing trend of interfacial energy with respect to the ionic liquid content.

The presence of increasing amounts of ionic liquid tended to improve the spreading of the water drop over the film surface resulting in an improvement of surface hydrophilic character due to the presence of ionic liquid on the surface of MWCNT. As the ionic liquid concentration increases, S_c is shifted to lower negative values as compared composites containing unmodified MWCNT which further confirms the hydrophilicity of prepared composites with increase in ionic liquid loading. In conclusion, the spreading coefficient seems to be affected mainly by two factors, the chemical composition (i.e., the amount of hydrophilic atoms) and the surface morphology of the composites.[22] On examining the Girifalco–Good's interaction parameter, it is found that its value increases with increase in ionic liquid concentration suggesting the better interaction between polymer surface and test liquid with ionic liquid concentration.

3.4 CONCLUSION

- For both the solvents water and DMSO, SBR nanocomposites show increase in contact angle with increase in f-MWCNT concentration up to a certain filler loading.
- Nanocomposites containing f-MWCNT (1:1 modification) exhibit nonwetting behavior.
- From various parameters such as surface energy, work of adhesion, polarity, spreading coefficient, interfacial energy, and interaction parameter of all composites, it is well evident that water is a nonwetting agent than DMSO.
- Increasing the concentration of surface modifier resulted in a transition from nonwetting to wetting behavior due to the surface coverage of MWCNT by ionic liquid.
- Tunable surface characteristics are obtained by controlling the chemical structure and composition of the polymer surfaces.

ACKNOWLEDGMENT

Financial support from Council for Scientific and Industrial Research (CSIR), Delhi, India is greatly acknowledged.

KEYWORDS

- styrene-butadiene rubber
- multiwalled carbon nanotube
- ionic liquid
- contact angle
- wetting behavior
- hydrophilicity
- hydrophobicity

REFERENCES

1. Zeng, Q. H.; Yu, A. B.; Lu, G. Q.; Paul, D. R. *J. Nanosci. Nanotechnol.* **2005,** *5* (10), 1574–1592.
2. Hameed, N.; Thomas, S. P.; Abraham, R.; Thomas, S. *Express Polym. Lett.* **2007,** *1,* 345–355.
3. Tai, Y. L.; Qian, J. S.; Miao, J. B.; Xia, R.; Zhang, Y. C.; Yang, Z. G. *Mater. Des.* **2012,** *34,* 522–527.
4. Vatanpour, V.; Madaeni, S. S.; Moradian, R.; Zinadini, S.; Astinchap, B. *Sep. Purif. Technol.* **2012,** *90,* 69–82.
5. Zhang, Y.; Zhang, Q.; Liu, Q.; Cheng, H.; Frost, R. L. *J. Therm. Anal. Calorim.* **2014,** *115* (2), 1013–1020.
6. Jose, J. P.; Abraham, J.; Maria, H. J.; Varughese, K. T.; Thomas, S. *Macromol. Symp.* **2016,** *366,* 66–78.
7. Young, T. *Philos. Trans. R. Soc. Lond.* **1805,** *95,* 65–87.
8. Fox, H.; Zisman, H. W. *J. Colloid Sci.* **1952,** *7,* 428–442.
9. Fowkes, F. M. *J. Phys. Chem.* **1963,** *67,* 2538–2541.
10. Owens, D. K.; Wendt, R. C. *J. Appl. Polym. Sci.* **1969,** *13,* 1741–1747.
11. Kaelble, D. H. *J. Adhes.* **1970,** *2* (2), 66–81.
12. Wu, S. *Polymer Interface and Adhesion*; M. Dekker, 1982.
13. Esposito, L. H.; Ramos, J. A., Kortaberria, G. *Prog. Org. Coat.* **2014,** *77* (9), 1452–1458.
14. Natarajan, B.; Li, Y.; Deng, H.; Brinson, L. C.; Schadler, L. S. *Macromolecules* **2013,** *46,* 2833–2841.
15. Varughese, K. T.; De, P. P.; Sanyal, S. K. *J. Adhes. Sci. Technol.* **1989,** *3,* 541.
16. Babu, S. S.; Mathew, S.; Kalarikkal, N., Thomas, S. *3 Biotech* **2016,** *6* (2), 249.
17. Zhang, F. H.; Wang, R. G.; He, X. D.; Wang, C., Ren, L. N. *J. Mater. Sci.* **2009,** *44* (13), 3574–3577.
18. Li, M.; Gu, Y.; Liu, Y.; Li, Y., Zhang, Z. *Carbon* **2013,** *52,* 109–121.
19. Phao, N.; Nxumalo, E. N.; Mamba, B. B., Mhlanga, S. D. *Phys. Chem. Earth Parts A/B/C* **2003,** *66,* 148–156.
20. Milionis, A.; Languasco, J.; Loth, E., Bayer, I. S. *Chem. Eng. J.* **2015,** *281,* 730–738.
21. Kotsalis, E. M.; Demosthenous, E.; Walther, J. H.; Kassinos, S. C., Koumoutsakos, P. *Chem. Phys. Lett.* **2005,** *412* (4), 250–254.
22. Armentano, I.; Fortunati, E.; Gigli, M.; Luzi, F.; Trotta, R.; Bicchi, I., Torre, L. *Polym. Degrad. Stab.* **2016,** *132,* 220–230.

CHAPTER 4

PROGRESS ON CARBON NANOTUBE PULL-OUT SIMULATION WITH PARTICULAR APPLICATION ON POLYMER MATRIX VIA FINITE ELEMENT MODEL METHOD

M. ESMAEILI, R. ANSARI, and A. K. HAGHI*

Faculty of Engineering, University of Guilan, Rasht, Iran

Corresponding author. E-mail: akhaghi@yahoo.com

ABSTRACT

The present study investigates the interfacial properties of carbon nanotube (CNT)-reinforced polymer composites by simulating nanotube pull-out. A finite element model of the noncovalent van der Waals interaction between CNTs and polymer via pull-out method has been developed to evaluation of the interfacial properties of CNT–polymer interphase. The Lennard-Jones interatomic potential was employed to simulate a nonbonded interphase. In this method, the interfacial shear stress during displacement of CNT has been determined directly through calculating shear force on surface of each carbon atom during pull-out. The shear stress distribution along CNT length for better understanding of pull-out results has been determined as novelty of this method. This model successfully obtains the maximum pull-out force to calculate interfacial shear strength (ISS). The research findings contribute to a better understanding of the load transfer on the tube–polymer interphase through CNT and the tube's reinforcing mechanism. Based on the pull-out modeling technique, the effects of CNT length on pull-out force, ISS and stress have also been analyzed in details. The results indicated that, with the increase in the embedded length of nanotube, the interface debonding process leads to an unchanged debonding force when the embedded length exceeds a threshold value named as "critical embedded length"; also, the ISS has been decreased with increase in CNT length. In this method clearly difference between catastrophic and noncatastrophic CNT debonding during pull-out with change in the embedded length of CNT have been investigated.

4.1   INTRODUCTION

Since the discovery of carbon nanotubes (CNTs) in 1991,[1] it has attracted much attention because of their unique structure and outstanding mechanical, electrical, thermal, and chemical properties.[2–4] Because of superior mechanical properties of CNTs, it was emerged as the perfect reinforcements for various applications in advanced nanocomposites.[5–9] It has been understood that the mechanical properties of CNT-reinforced polymer composites are dominated mainly by interfacial shear strength (ISS) between CNTs and the matrix and length of CNT.[5–7] The load transfer between CNT and the matrix plays an important role in the mechanical properties of the nanocomposite, so it is a complex process that depends on the interfacial bonding between the CNT reinforcement and the surrounding polymer matrix. The stress transfer between the CNTs and the polymer matrix at the interface is a key factor which strongly affects the mechanical properties of CNT-reinforced polymer composites. The evidence of such stress transfer from the external forces to the CNTs has been reported by some researchers.[5,8] In this regards, different methods have been used in the analysis of the stress transfer between CNT and polymer matrix. Fiber pull-out tests have been recognized as the standard method for evaluating the interfacial characteristics of composite materials. The pull-out experiments of individual CNTs from the polymer matrix have been carried out extremely by Wagner group.[10–14] They measured the force required to separate a CNT from a solid polymer matrix using atomic force microscopy. They successfully traced the pull-out force and nanotube locations to obtain the force–displacement curve and average ISS and demonstrated that the ISS between a multiwall carbon nanotube (MWCNT) and an epoxy matrix is in the range of 35–376 MPa using a scanning probe microscope (SPM) setup to drag the nanotube out from the matrix.[11] Later, they employed an atomic force microscope (AFM) to directly pull a MWCNT from a polyethylene–butene matrix and observed an ISS of 47 MPa[12] and the experimental data reported on the CNT–epoxy interface was 30 ± 7 MPa.[14] Their experimental results exhibit high pull-out forces at the initial stage of the pull-out process. This is immediately followed by a sudden decay in the force until the CNT has fully been withdrawn from the polymer. The pull-out of the CNT from the polymer at first involves an increase in the pull-out force while the nanotube remains in full contact with the polymer.[12] More displacement of the nanotube causes initiation and growth of a debonded region until nanotube fully debonds from the polymer matrix at a maximum force. The full debonding is followed by a sharp drop in the recorded force due to a fully failed interface.[13] Their measurements for higher embedded lengths shown that the force required to pull each nanotube out of the polymer is seen to increase as the embedded length increases.[14] Chen et al.[15] presented a direct measurements of the interfacial stress transfer between polymers [poly(methyl methacrylate) (PMMA)] and CNTs of sub-10 nm in outer diameter (that remained largely unexplored) using an in situ nanomechanical single-tube pull-out testing scheme inside a high-resolution electron microscope. Their results indicated that the CNT–PMMA interface possesses ISS within 32–68 MPa. The SEM observation revealed that the pull-out of the nanotube occurred as a catastrophic failure of the CNT–polymer interface. By pulling out individual tubes with different embedded lengths, their work reveals the shear lag effect on the nanotube–polymer interface and demonstrates that the effective interfacial load transfer occurs only within a certain critical embedded length. Unfortunately, some differences has been observed in the above various experimental data, largely due to the difficulty in trustworthy of measuring the pull-out force or strain at the nanoscale specially in short length of CNT. Anyway, numerical simulation such as continuum mechanics, molecular mechanics (MM), and molecular dynamics (MD) confirmed by recent numerical investigation into the interfacial behavior of CNT/polymer nanocomposites.

Zheng et al.[16] have reported the ISS of 33 MPa for nanotubes in a polyethylene matrix using MD. Also, the MD model of Xu et al.[17] predicted shear strength around 138 MPa, considering nonbonded interactions between a nanotube and epoxy matrix. Li and Chou[18] developed a computational model to evaluate the ISS in nanotube reinforced composites. Wernik et al.[19] evaluated the interfacial properties of CNT-reinforced polymer composites using atomistic-based continuum model. Meguid et al.[20] conducted MD simulations to determine the atomic-level interface. However, the length of the CNT in the MD models was limited to the range of 4–10 nm due to the intensive computational requirements in the MD simulations. In the MD pull-out studies,[21–24] the force required to withdraw the CNT from the matrix is evaluated over the course of the pull-out process by summing the reaction forces at the upper CNT nodes. The corresponding ISS can then be calculated by dividing the maximum pull-out force by the initial interfacial area. Their results shown that the increasing the CNT-embedded length has no effect on the maximum pull-out force of nonbonded nanotubes; rather it only serves to extend the sliding regime of the pull-out profile. Pull-out force/displacement figures show that different lengths have the same amount of pull-out force; the pull-out forces calculated by MD simulations seem to increase up to a relatively fixed value until the last stages of a pull-out, as well as complete pull-out of the nanotube happened at the full-length displacement. Based on van der Waals (vdW) interaction that considered at the interface using the Lennard-Jones (LJ) potential, with increase in CNT displacement in pull-out, amount of vdW interactions will be decreased because of loss in efficient distance between CNT and polymer atoms and consequently it cannot be reasonable having constant pulling force until complete removal of CNT from polymer matrix. This issue was not considered in pull-out diagram in MD simulations. Although MD can describe interactions at atomic levels through suitable potential models, it is limited by length and time scales due to the small time steps required. Some conclusions that are observed from experimental studies have not been seen in numerical and analytical researches. Specifically, high pull-out forces have been shown at the initial stage of a pull-out process in experimental studies. A sudden decay in the force has been observed until a CNT has fully been pulled out from the polymer. On the contrary, the pull-out forces predicted by numerical simulations seem to increase up to a relatively fixed value until the last stages of a pull-out process. Therefore, several fundamental differences between experimental and numerical studies are worthy of investigation. Although MD can describe interactions at atomic levels through suitable potential models, it is limited by length and time scales due to the small time steps required. These limitations make continuum mechanics approaches more favorable for analyses at length scales in the micron range. Continuum mechanics-based approaches developed include cohesive zone model,[25–28] Cox's model,[29] and shear-lag model.[30–32] Viet and Kuo[33] employed both shear-lag modeling and finite element method to study the load transfer in fractured CNTs. Tsai and Lu[34] applied the equivalent continuum geometry for the three-walled CNT-reinforced composite to study the load transfer from matrix to three interior walls using a shear-lag model and the finite element analysis. Haque and Ramasetty[35] studied the axial and shear stress at the interface of CNT-reinforced polymer composite materials. They investigated the efficiency of load transfer in CNT-reinforced polymer for the effective length of the CNT. A micromechanics model has also been developed by Li and Saigal[36] for assessing the ISS transfer in CNT-reinforced polymer composites. Their results indicated that the stress transfer characteristics of nanocomposites can be improved by sufficiently long CNTs. The classical shear-lag model has been adopted to predict the interfacial stress transfer in CNT-reinforced polymer composites.[37] Nonlinear cohesive laws for CNT–polymer matrix interfaces have been developed based on the vdW interfacial interaction chemical bonding to analyze the CNT pull-out from a polymer matrix.[38,39] Also, cohesive zone finite element models (FE) have been established with introducing a nonlinear interface cohesive law to model the pull-out response of CNT-coated fibers.[40,41] Recently, the finite element method was applied to simulate the single CNT pull-out with the assistance of the cohesive zone model to understand its bridging effect in CNT-reinforced composites.[42] A FE with the assistance of the cohesive zone model was applied to simulate a single CNT pull-out from a polymeric matrix using cohesive zone modeling by Yan et al.,[28] their numerical results indicate that the debonding force during the CNT pull-out increases almost linearly with the interfacial crack initiation shear stress and a saturated debonding force exists corresponding to a critical CNT-embedded length. Chen and Yan[32] used different cohesive laws together with the classical shear-lag theory to establish an analytical relationship between the load transfer and the interfacial properties between CNT and matrix. Analytical expressions for the maximum fiber pull-out force and its limit as the embedded fiber length approaches infinity was obtained. The continuum mechanics approaches method cannot cover all assumptions in nanoscale and correlation between nano- and microlevels. So, their calculations show overestimate results.

The limitations in such theoretical and experimental methods for characterizing the CNT–matrix interfacial properties, and the lack of a complete method for full study of different (include short and long) length effects on stress transfer properties in CNT nanocomposite, inspire us to establish a method for determining interfacial properties and study of length effects. We first derive the finite element method for model a representative volume element (RVE) consisting of CNT, polymer matrix and fiber–matrix interphase by introducing a nonlinear vdW interaction between CNT and matrix. The main contribution of this research is direct measurement of pull-out force in interphase that caused by vdW interactions during different displacement. Consequently ISS calculated from the pull-out force. Measurement of shear stress distribution across the nanotube length and availability of shear stress measuring in different lengths, diameters, and chirality of nanotubes are another advantages of this method. By pulling out CNT with different wide range of embedded lengths, from 7.2707 to 153.9 nm, shear stress/displacement diagram will be studied. This study demonstrates that our FE can be used to characterize the interfacial properties, thus enabling a convincing quantification and comparison of the interfacial strength across different nanotube length.

4.2 MODEL PROCEDURE

A FE was employed in this study to simulate RVE that consists of three separate regions as: CNT, interphase, and surrounding polymer. To simulate CNT, based on chirality of CNT, positions of carbon atoms are determined. To simulate carbon bond (C–C), a beam element according to the Lee and Chou[18] was used. They established a correlation between the interatomic molecular potential energy and the strain energy of the beam. Using this method, elastic modulus (E) and Poisson coefficient (v) per carbon–carbon bond are obtained. The CNT is modeled as a space-frame structure as depicted in Figure 4.1.

In the space-frame model, each beam element corresponds to an individual chemical bond in the CNT. As in traditional FE, nodes are used to connect the beam elements to form the CNT structure. In this case, the nodes represent the carbon atoms and their positions are defined by the same atomic coordinates. Young's modulus of the CNT model was obtained 1.068 MPa and has a good agreement with data published in the.[43] The polymer around CNT can be modeled as a continuums medium, which is an acceptable simplification in the modeling procedure.[43] Resin with specific properties (E, v) can be modeled using three-dimensional solid elements. vdW forces between atoms of carbon and resin are weak forces. Their role is force transmission between the carbon atoms and resin and the distance between the atoms of carbon and resin determines the amount of the vdW force. vdW bonds between the carbon atoms and resin, created by a developed code, this program reads the position of the atoms of carbon and resin and calculates the distance between them (D). The program has been developed capable of remeshing, this means that after the apply boundary conditions on the end of resin and special displacement of CNT to axial direction that illustrated in Figure 4.2. Some vdW interactions are deactivated when their length exceeds the cutoff distance dictated by the LJ potential (>0.85 nm) according to Figure 4.2, some new vdW links will be generated if interatomic distance is less than or equal to 0.85 nm accordance with the new situation of the model (Fig. 4.3).

The vdW forces, which are nonlinear, are modeled by using LJ "6–12" as mentioned in eq 4.1[44]:

$$F_{vdw} = 4\frac{\varepsilon}{r}\left[-12\left(\frac{\sigma}{r}\right)^{12} + 6\left(\frac{\sigma}{r}\right)^{6}\right] \tag{4.1}$$

ε and σ are the LJ parameters, equal to 0.4492 kJ/mol and 0.3825 nm, respectively.[45]

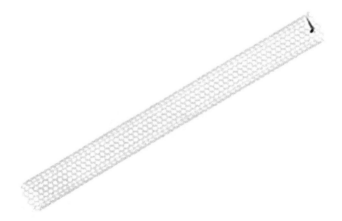

FIGURE 4.1   Space-frame structure of CNT.

So the process has a number of bonds lost and another is created. After all the above steps and solving the problem completely for certain displacement, the forces due to movement of nonlinear springs to the surface of the carbon atoms, in the axial direction will be extracted. By dividing the total force on the surface of the nanotubes, shear stress generated in the interphase can be calculated. Figure 4.4 shows the chart of modeling procedure.

4.3   FINITE ELEMENT METHOD

In this study, an APDL code (ANSYS program design language) in commercial finite element software ANSYS has been developed. In this model each C–C bond was modeled by BEAM188. The element is based on Timoshenko beam theory which includes shear-deformation effects. The element is a linear, quadratic, or cubic two-node beam element in 3-D. BEAM188 has six degrees of freedom at each node. These include translations in the x, y, and z directions and rotations about the x, y, and z directions. The polymer resin was modeled by 3-D SOLID95 element. This element has three degrees of freedom per each node, in the x, y, and z directions, and can tolerate irregular shapes without the loss of accuracy. The vdW interactions between the carbon atoms of CNT and nodes of the inner surface of the resin were modeled using a nonlinear 3-D spring. A COMBIN39 element was employed in the ANSYS software for this

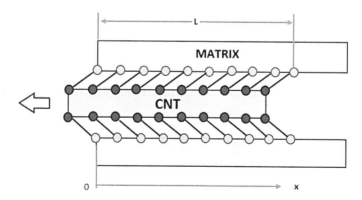

FIGURE 4.2   CNT displacement trough axial direction.

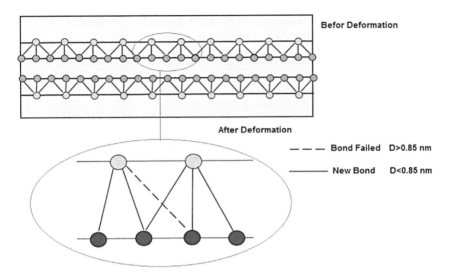

FIGURE 4.3 Debonding and new bonding of vdW interaction during displacement of CNT.

purpose. COMBIN39 is a unidirectional element with nonlinear generalized force-deflection capability that can be used in any analysis. The longitudinal option is a uniaxial tension–compression element with up to three degrees of freedom at each node: translations in the nodal x, y, and z directions.[46] An APDL code was developed to model the interphase region, which created nonlinear spring elements between CNT atoms and inner surface atoms of the polymer, whose distances was less than or equal to 0.85 nm. The properties of the nonlinear spring elements have been calculated by eq 4.1. The thickness of CNT was selected to be 0.34 nm, and the centers of carbon atoms in the CNT were placed at the midsection of tube thickness. So the equilibrium distance between carbon and polymer atoms is 0.17 nm.[43,47] All three parts of the FE consisting of a CNT, the interphase, and the surrounding polymer are shown in Figure 4.5. Also, different pictures of CNT pull-out process are illustrated in Figure 4.6.

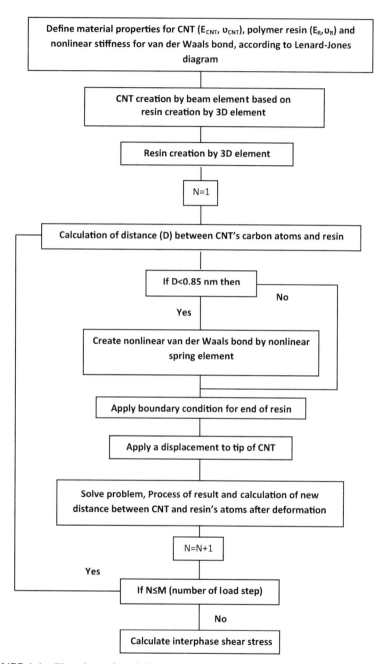

FIGURE 4.4 Flowchart of modeling.

4.4   MODEL VALIDATION

The CNT selected in this study for validation of the model is an armchair single-walled CNT (SWCNT) with a chiral index (10, 10) SWCNT and uncapped, with a length of 170 nm.[15] The geometry of RVE is presented in Table 4.1.

A pull-out displacement is applied to the tip of the CNT in the axial direction. The ISS of SWCNT–polymer is evaluated at interfaces as[48,49]:

$$\tau_s = \frac{F_{pull\text{-}out}}{\pi \times l_{CNT} \times D} \tag{4.2}$$

l_{CNT} is embedded CNT length, D is CNT diameter and $F_{pull\ out}$ is the pull-out force at interfacial rupture, which is one of the most important resulted parameters from a pull-out test. The calculated ISS (33.358 MPa) has good agreement with reported experimental afford data on the CNT–epoxy interface (30 ± 7 MPa),[14] short embedded lengths less than 310 nm (32–68 MPa),[15] and theoretically predicted values of 27.4–35.9 MPa for the interface formed by SWCNTs with PMMA based on MD simulations.[16] Comparison results of ISS for different calculation methods were shown in Table 4.2.

TABLE 4.1   Geometrical Properties of Validation RVE.

Geometrical parameters	Quantity	Unit
Length of CNT (L)	170 [15]	nm
Length of resin	170 [15]	nm
Diameter of CNT (D)	2 [15]	nm
Thickness of CNT (t_{CNT})	0.34 [43, 47]	nm
Thickness of resin (t_r)	2.1968 [43, 47]	nm
Distance between resin and CNT (t_{ir})	0.17 [43, 47]	nm
Carbon–carbon bond length	0.142 [43, 47]	nm
Carbon–carbon bond modulus	5.78	TPa
Poisson efficiency of carbon–carbon bond	0.3 [43, 47]	–
Elasticity's modulus of resin	2 [15]	GPa
Poisson coefficient of resin	0.32 [15]	–

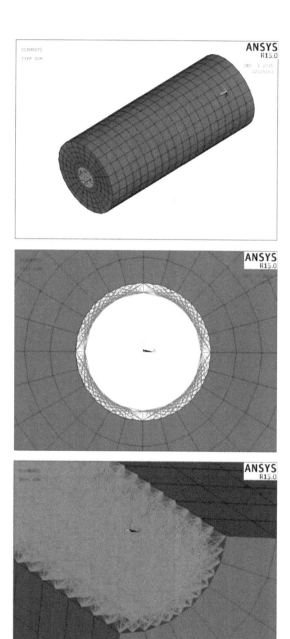

FIGURE 4.5 **(See color insert.)** Three parts of RVE from different point of view.

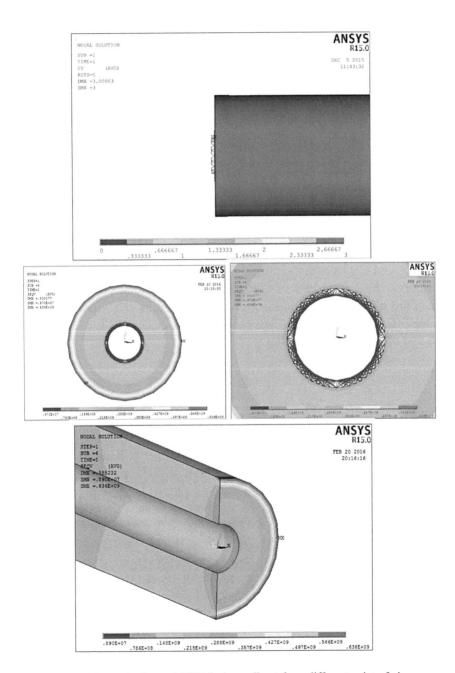

FIGURE 4.6   (See color insert.) RVE during pull-out from different point of view.

TABLE 4.2 Comparison of ISS for SWCNTs.

Diameter (nm)	Measurement method	ISS	References
–	MD	27.4–35.9	[16]
2–4.2	Experimental	32–68	[15]
–	Experimental	30 ± 7	[14]
2	FE	33.538	Present study

FE, finite element model; ISS, interfacial shear strength; MD, molecular dynamics.

4.5 CNT PULL-OUT

4.5.1 INTERFACIAL SHEAR STRESS DIAGRAM

The ISS has been caused on the CNT–polymer interphase in response to the applied displacement on tip of the CNT, in pulling out from the matrix due to displacement in nonlinear springs. The value of the shear stress is actually distributed nonuniformly across the entire CNT–polymer interphase. The interfacial shear stress for each displacement is calculated for the whole interfacial area that is given by:

$$\tau = \frac{F_{total}}{\pi \times l_{CNT} \times D} \tag{4.3}$$

F_{total} is the total tangential force that is calculated on the surface of carbon atoms in CNT for each substep pull-out displacement that caused by springs displacement. Figure 4.7a shows the calculated interfacial shear stress versus pull-out displacement in the CNT–polymer interphase, the SWCNT (10, 10) has been used with $D = 1.356$ nm and embedded length of $L = 29.412$ nm. Three different steps have been observed in the shear stress-displacement diagram (Fig. 4.7b). At the first step, with increasing in displacement the shear stress is increased linearly and then linear behavior terminates at displacement = 0.9 nm and $\tau = 168.5$ MPa. In this area that called elastic area, the shear stress-displacement diagram shows a linear behavior that corresponded to the elastic loading with fully bonded interphase, and means no damage occurred at vdW bonding in the interphase area during pull-out. In the second area of diagram, shear stress increases nonlinearly and reaches to the maximum amount (displacement = 2.2 nm, $\tau_{max} = 239.5$ MPa), in next steps of displacement. This area called damage area that means some vdW bonds in the interphase have been damaged during displacement but the total lost bonds are not enough for start debonding. After gaining τ_{max}, declining in shear stress will be started and this means the debonding will be started at the maximum interfacial stress. Hence, after initial debonding, further debonding requires to applying stress to overcome the interfacial shear stress at the bonded interphase. Consequently, the stress required for complete debonding depends on the extent of prior debonding. The complete debonding occurs at displacement = 2.82 nm, τ = 216 MPa and after that, the complete pull-out will occur. It should be considered that the interfacial shear stress due to friction is assumed to be zero in these calculations.

4.5.2 DISTRIBUTION OF INTERFACIAL SHEAR STRESS ALONG CNT LENGTH

One of the most important novelties in this research is calculation of shear stress distribution along the CNT to investigate different pull-out phases more carefully. The shear stress distribution along CNT is illustrated in Figures 4.8–4.11 for three pull-out steps that mentioned in Section 4.5.1. Figure 4.8 shows shear stress distribution in elastic area for different displacements. In the first displacement (0.5 nm) the maximum value of shear stress occurred at the end of CNT entry depth to the polymer position. All CNT's length participated in stress transfer; the maximum amount of shear stress is reached at the end of CNT, and with more displacement local maximum shear stress will be increased. Shear stress distribution at

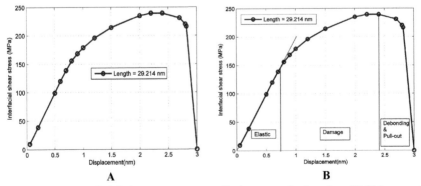

FIGURE 4.7 Interfacial shear stress versus displacement for length = 29.214 nm and different steps of pull-out.

distribution of 0.9 nm in slightly changed form can be seen, that can indicate the beginning of a phase change in stress transfer behavior.

Figure 4.9 shows the shear stress distribution along CNT at displacement of 0.9 nm (the end of elastic area) and displacement of 1 nm (beginning of damage area). As seen in a length of about 3 nm (from the side of entry in polymer of nanotubes), the amount of shear stress in the diagram of the elastic area is greater than the value in the graph corresponding to the area of damage. In fact, by the damage and the loss of part of vdW interaction shear stress is reduced in that area. With the increase in displacement, more interactions will be lost and damage will be developed.

Distribution of shear stress in damage area is shown in Figure 4.9 for different displacements. With increase in displacement, the local maximum shear stress along the CNT is shifted from CNT entry depth to the polymer to the other side of embedded length of CNT. These graphs show well that with increasing displacement, the maximum local shear stress is increased

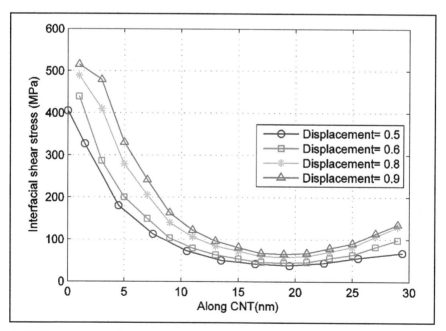

FIGURE 4.8   **(See color insert.)** Shear stress distribution along CNT's length in elastic area.

and reaches a constant value around 500 MPa. As shown in Figure 4.7, by more increases in the displacement, the local maximum shear stress remained constant, but the amount of shear stress, until reaching the maximum amount is reduced, which represents the increase in damage.

Figure 4.11 demonstrated the shear stress distribution at debonding area. As can be seen from Figure 4.11, after displacement 2.2 nm and achieve maximum shear stress, debonding is occurred in a portion of the nanotubes. The debonding length increases with increase in displacement until the maximum debonded length of CNT at 5 nm for displacement 2.8 nm; after that the complete debonding and pull-out occur at displacement 2.82 nm.

All of these three figures illustrate local shear stress distribution for each step of pull-out process. It is clearly indicated that all of the diagrams have a local maximum amount that is near to left end of CNT in elastic pull-out area and right end in debonding area, and it will be shifted from left to right in damage zone in more displacement.

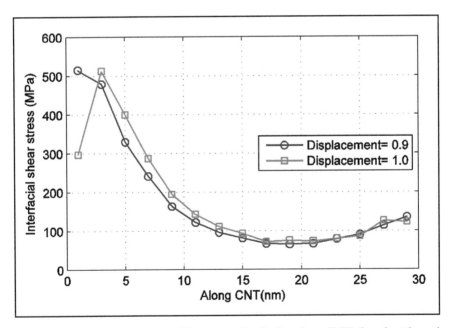

FIGURE 4.9 **(See color insert.)** Shear stress distribution along CNT's length at the end of elastic area (displacement 0.9 nm) and initial damage area (displacement 1 nm).

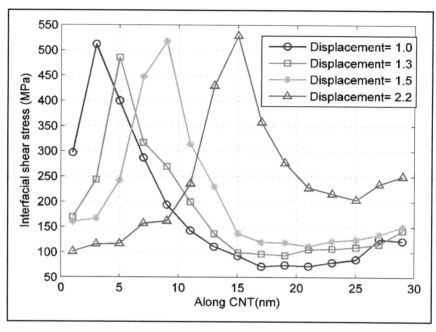

FIGURE 4.10   **(See color insert.)** Shear stress distribution along CNT's length in damage area.

4.5.3   EFFECT OF CNT LENGTH

The CNT length is one of the most important parameters in the properties of the interphase between CNT and surrounding polymer matrix. To better understand the effect of the CNTs length, different CNT lengths were evaluated and their effects on shear properties were investigated. Figure 4.6 shows shear stress-displacement diagram for three different lengths. As we can see in Figure 4.6 with decrease in CNT length, the elastic area has become more sharp, damage and debonding zone will occur in shorter area and the maximum interfacial shear stress will be increased.

In short CNTs, the drop in shear stress from the maximum amount is usually followed by a pull-out shoulder after complete debonding, and pull-out will be *noncatastrophic* for (7.2707 and 14.644 nm) in comparison with *catastrophic* in longer CNT length (29.214 nm) as shown in Figure 4.7. The *noncatastrophic* pull-out area is clearly shown in Figure 4.7.

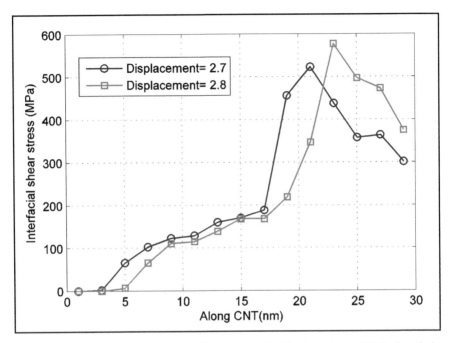

FIGURE 4.11   (See color insert.) Shear stress distribution along CNT's length in debonding area.

The noncatastrophic pull-out in traditional polymer composite reflects the force required overcoming the frictional interaction between fiber and polymer matrix after debonding,[50] but as CNTs are atomically smooth, it is reasonable to assume that the nanotube slides out relatively easily from the surrounding polymer matrix after full nanotube debonding. It is essential to differentiate clearly between catastrophic and noncatastrophic debonding in order to determine the true shear strength of the interface. With increase in CNT's length, the interface region is saturated with a high number of vdW interactions which prevents any sharp changes in the pull-out force as the CNT is withdrawn from the matrix until complete debonding (Fig. 4.14).

Figure 4.15 shows the interfacial shear stress-displacement in different embedded length of CNT ranging from 7.2707 to 153.9 nm. Our results show that the maximum shear stress decreases with increase in embedded length.

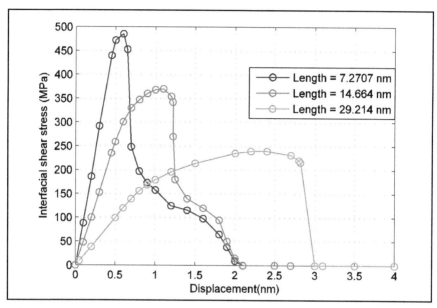

FIGURE 4.12 (**See color insert.**) Shear stress diagram for different length shorter than 29.412 nm.

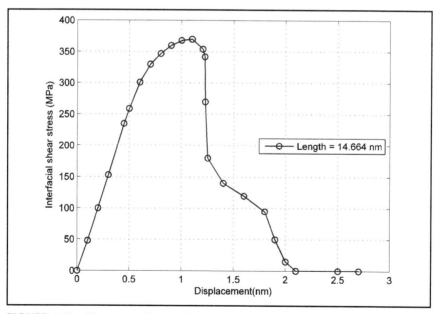

FIGURE 4.13 Shear stress diagram for length 14.644 nm.

4.5.4 SATURATED DEBONDING FORCE AND CRITICAL CNT-EMBEDDED LENGTH

Figure 4.16 shows the pull-out forces diagram for different displacements measured in embedded nanotube length ranging from 7.2707 to 153.9 nm. The results show that the maximum pull-out force that is called debonding force, first increases with the increase in embedded length (up to 29.214 nm), and then approaches a saturated value when the embedded length of CNT reaches a critical value.

The interface debonding process leads to an unchanged debonding force when the embedded length exceeds a threshold value named as critical embedded length. As shown in Figure 4.17, the debonding force becomes saturated (F_{sat} = 30 nN) and the critical CNT-embedded length is about 29 nm. When $L_{CNT} < L_{c \ (CNT)}$, the debonding force increases with increasing CNT-embedded length (L_{CNT}). When $L_{CNT} = L_{c \ (CNT)}$

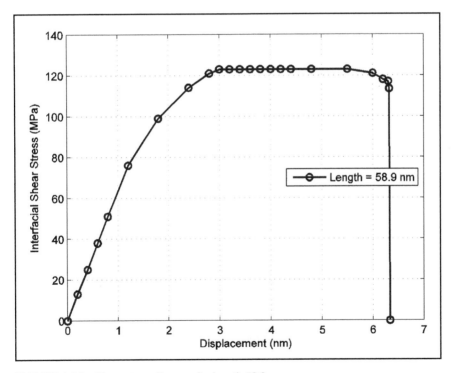

FIGURE 4.14   Shear stress diagram for length 58.9 nm.

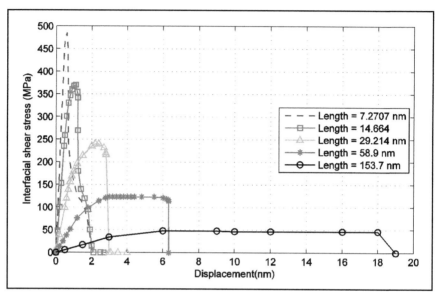

FIGURE 4.15 **(See color insert.)** Shear stress diagram for different length from 7.2707 to 153.7 nm.

or $L_{CNT} > L_{c\ (CNT)}$ the debonding force becomes saturated. This numerical result indicates that the debonding force cannot be continuously increased by increasing the CNT-embedded length, and the increasing is only effective for short CNTs.

4.6 CONCLUSION

In this chapter, we present a study of finite element method of the CNT–polymer interfacial stress transfer using pull-out method. The research findings presented in this paper contribute to a better understanding of dependence of interfacial shear stress to pull-out displacement, maximum shear stress and clearly distribution of shear stress. The results demonstrated that the average interfacial shear stress is increased with increase in pull-out displacement to a maximum value. Furthermore, the interfacial shear stress-displacement diagram exhibited three distinguished phases during pull-out process; elastic or linear area, damage area, and debonding

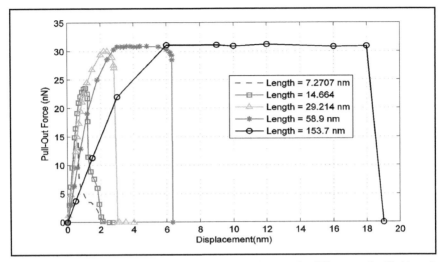

FIGURE 4.16 **(See color insert.)** Pull-out forces diagram for different embedded length ranging from; saturated value of maximum pull-out force.

and pull-out area. The shear stress distribution along CNT lengths clearly confirmed these three phase.

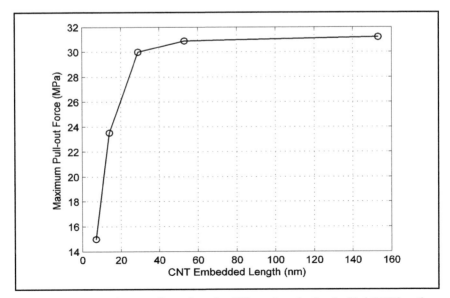

FIGURE 4.17 Maximum pull-out force for different length of embedded CNT length.

At the separate parts in this research, we present a study on interfacial stress of the CNT–polymer nanocomposite using an FE for simulation nanotube pull-out test technique. This method clearly proposed difference between catastrophic and noncatastrophic CNT debonding during pull-out with change in embedded CNT length. Our measurements of the pull-out force for CNTs with different embedded lengths clearly reveal that the effective shear load transfer on the nanotube–polymer interphase is confined within the critical embedded length. Moreover, our analysis highlights that it is essential to take into account the critical embedded length in the evaluation of the nanotube–polymer interfacial stress.

KEYWORDS

- **carbon nanotubes**
- **pull-out**
- **interfacial shear strength**
- **finite element analysis**
- **critical length**
- **maximum pull-out force**
- **debonding force**

REFERENCES

1. Iijima, S. Helical Micro Tubes of Graphic Carbon. *Nature* **1991**, *354*, 56–58.
2. Demcayk, B. G.; Wang, Y. M.; Cumings, J.; Hetman, M.; Han, W.; Zettl, A. Direct Mechanical Measurement of the Tensile Strength and Elastic Modulus of Multi-walled Carbon Nanotubes. *Mater. Sci. Eng. A* **2002**, *334*, 173–178.
3. Maruyama, S. A. Molecular Dynamics Simulation of Heat Conduction of Finite Length SWNTs. *Physica. B* **2002**, *3* (23), 193–195.
4. Maruyama, S. A Molecular Dynamics Simulation of Heat Conduction of a Finite Length Single-walled Carbon Nanotube. *Micro Thermophys. Eng.* **2003**, *7*, 41–50.
5. Teacy, M. M. J.; Ebbesen, T. W.; Gibson, J. M. Exceptionally High Young's Modulus Observed for Individual Carbon Nanotubes. *Nature* **1996**, *381* (6584), 678–680.
6. Yakobson, B. I.; Brabec, C. J.; Bernholc, J. Nanomechanics of Carbon Tubes: Instabilities Beyond Linear Response. *Phys. Rev. Lett.* **1996**, *76*, 2511.

7. Latibari, S. T.; Mehrali, M.; Mottahedin, L.; Fereidoon, A.; Metselaar, H. S. C. Investigation of Interfacial Damping Nanotube-based Composite. *Composites Part B* **2013,** *50,* 354–361.

8. Kundalwal, S. I.; Ray, M. C. Effect of Carbon Nanotube Waviness on the Effective Thermoelastic Properties of a Novel Continuous Fuzzy Fiber Reinforced Composite. *Composites Part B* **2014,** *57,* 199–209.

9. Kundalwal, S. I.; Ray, M. C.; Meguid, S. A. Shear Lag Model for Regularly Staggered Short Fuzzy Fiber Reinforced Composite. *ASME J. Appl. Mech.* **2014,** *81* (9) 091001.

10. Lourie, O.; Wagner, H. D. Evidence of Stress Transfer and Formation of Fracture Cluster in Carbon Nanotube-based Composites. *Compos. Sci. Technol.* **2010,** *59,* 975–977.

11. Cooper, C. A.; Cohen, S. R.; Barber, A. H.; Wagner, H. D. Detachment of Nanotubes from a Polymer Matrix. *Appl. Phys. Lett.* **2002,** *81,* 3873–3875.

12. Barber, A. H.; Cohen, S. R.; Wagner, H. D. Measurement of Carbon Nanotube Polymer Interfacial Strength. *Appl. Phys. Lett.* **2003,** *82,* 41–40.

13. Barber, A. H.; Cohen, S. R.; Kenig, S.; Wagner, H. D. Interfacial Fracture Energy Measurements for Multi-walled Carbon Nanotubes Pulled from a Polymer Matrix. *Compos. Sci. Technol.* **2004,** *64,* 2283–2289.

14. Barber, A. H.; Cohen, S. R.; Eitan, A.; Schadler, S. L.; Wagner, H. D. Fracture Transitions at a Carbon-nanotube/Polymer Interface. *Adv. Mater.* **2006,** *18,* 83–87.

15. Chen, X.; Zheng, M.; Park, Ch.; Ke, Ch. Direct Measurements of the Mechanical Strength of Carbon Nanotube–Poly(Methyl Methacrylate) Interfaces. *Small* **2013,** *19,* 3345–3351.

16. Zheng, Q.; Xue, Q.; Yan, Q. K.; Gao, X.; Li, Q.; Hao, L. Influence of Chirality on the Interfacial Bonding Characteristics of Carbon Nanotube Polymer Composites. *J. Appl. Phys.* **2008,** *103,* 044302.

17. Xu, X.; Thwe, M. M.; Hearwood, C.; Liao, K. Mechanical Properties and Interfacial Characteristics of Carbon-nanotube-reinforced Epoxy Thin Films. *Appl. Phys. Lett.* **2002,** *81,* 2833–2836.

18. Li, C.; Chou, T. W. Multiscale Modeling of Carbon Nanotube Reinforced Polymer Composite. *J. Nanosci. Nanotechnol.* **2003,** *3* (5) 423–430.

19. Wernik, J. M.; Cornwell-Mott, M. J.; Meguid, S. A. Determination of the Interfacial Properties of Carbon Nanotube Reinforced Polymer Composites Using Atomistic-based Continuum Model. *Int. J. Solids Struct.* **2012,** *49* 1852–1863.

20. Alian, A. R.; Kundalwal, S. L.; Meguid, S. A. Interfacial and Mechanical Properties of Epoxy Nanocomposites Using Different Multiscale Modeling Schemes. *Compos. Struct.* **2015,** *131,* 545–555.

21. Li, Y.; Hu, N.; Yamamoto, G.; Wang, Z.; Hashida, T.; Asanuma, H.; Dong, C.; Arai, M.; Fukunaga, H. Molecular Mechanics Simulation on Sliding Behavior Between Nested Walls in a Multi-walled Carbon Nanotube. *Carbon* **2010,** *48,* 2934–2940.

22. Meguid, S. A.; Wernik, J. M.; Cheng, Z. Q. Atomistic-based Continuum Representation of the Effective Properties of Nano-reinforced Epoxies. *Int. J. Solids Struct.* **2010,** *47,* 1723–1736.

23. Wernik, J. M.; Meguid, S. A. Multiscale Modeling of the Nonlinear Response of Nano-Reinforced Polymers. *Acta Mech.* **2011,** *217,* 1–16.
24. Xiong, Q. L.; Meguid, S. A. Atomistic Investigation of the Interfacial Mechanical Characteristics of Carbon Nanotube Reinforced Epoxy Composite. *Eur. Polym. J.* **2015,** *69,* 1–15.
25. Jiang, L. Y.; Huang, Y.; Jiang, H.; Ravichandran, G.; Gao, H.; Hwang, K. C.; Liu, B. A Cohesive Law for Carbon Nanotube/Polymer Interfaces Based on the van der Waals Force. *J. Mech. Phys. Solids* **2006,** *54,* 2436–2452.
26. Lu, W. B.; Wu, J.; Song, J.; Hwang, K. C.; Jiang, L. Y.; Huang, Y. A Cohesive Law for Interfaces Between Multi-Wall Carbon Nanotubes and Polymers due to the van der Waals Interactions. *Comput. Methods Appl. Mech. Eng.* **2008,** *197,* 3261–3267.
27. Jiang, L. Y. A Cohesive Law for Carbon Nanotube/Polymer Interface Accounting for Chemical Covalent Bonds. *Math. Mech. Solids* **2010,** *15,* 718–732.
28. Jia, Y.; Chen, Z.; Yan, W. A Numerical Study on Carbon Nanotube Pullout to Understand Its Bridging Effect in Carbon Nanotube Reinforced Composites. *Composite Part B* **2015,** *81* 64–71.
29. Xiao, K. Q.; Zhang, L. C. The Stress Transfer Efficiency of a Single-walled Carbon Nanotube in Epoxy Matrix. *J. Mater. Sci.* 2004, *39* (14) 4481–4486.
30. Gao, X. L.; Li, K. A Shear-Lag Model for Carbon Nanotube-reinforced Polymer Composites. *Int. J. Solids Struct.* **2005,** *42,* 1649–1667.
31. Keng, K.; Khondaker, A.; Ahmed, S. An Improved Shear-lag Model for Carbon Nanotube Reinforced Polymer Composites. *Composites Part B* **2013,** *50,* 7–14.
32. Chen, Z.; Yan, W. A Shear-lag Model with a Cohesive Fibre-matrix Interface for Analysis of Fiber Pull-out. *Mech. Mater.* **2015,** *91,* (1) 119–131.
33. Viet, N. V.; Kuo, W. S. Load Transfer in Fractured Carbon Nanotubes Under Tension. *Composites Part B* **2012,** *43,* 332–339.
34. Tsai, J. L.; Lu, T. C. Investigating the Load Transfer Efficiency in Carbon Nanotubes Reinforced Nanocomposites. *Compos. Struct.* **2009,** *90* (2), 172–179.
35. Haque, A. Ramasetty, A. Theoretical Study of Stress Transfer in Carbon Nanotube Reinforced Polymer Matrix Composites. *Compos. Struct.* **2005,** *71* 68–77.
36. Li, K.; Saigal, S. Micromechanical Modeling of Stress Transfer in Carbon Nanotube Reinforced Polymer Composites. *Mater. Sci. Eng. A* **2007,** *457,* 44–57.
37. Gao, X. L.; Li, K. A Shear-Lag Model for Carbon Nanotube-reinforced Polymer Composites. *Int. J. Solids Struct.* **2005,** *42,* 1649–1667.
38. Jiang, L. Y.; Huang, Y.; Jiang, H. G.; Ravichandran, H.; Gao, K. C.; Hwang, B.; Liu, A. Cohesive Law For Carbon Nanotube/Polymer Interfaces Based on the van der Waals Force. *J. Mech. Phys. Solids* **2006,** *5* (4) 2436–2452.
39. Lu, W.; Wu, B.; Song, J. K. C.; Hwang, J.; Jiang, L. Y.; Huang, Y. A Cohesive Law for Interfaces Between Multi-Wall Carbon Nanotubes and Polymers due to the van der Waals Interactions. *Comp. Methods Appl. Mech. Eng.* **2008,** *197,* 3261–3267.
40. Jiang, L. Y. A Cohesive Law for Carbon Nanotube/Polymer Interface Accounting for Chemical Covalent Bonds. *Math. Mech. Solids* **2010,** *15,* 718–732.

41. Agnihotri, P. K.; Kar, K. K.; Basu, S. Cohesive Zone Model of Carbon Nanotube-coated Carbon Fiber/Polyester Composites. *Model. Simul. Mater. Sci. Eng.* **2012**, *20*, 1–13.

42. Jia, Y; Chen, Z.; Yan, W. A Numerical Study on Carbon Nanotube-hybridized Carbon Fibre Pullout. *Compos. Sci. Technol.* **2014**, *91*, 38–44.

43. Shokrieh, M. M;. Rafiee, R. Prediction of Mechanical Properties of an Embedded Carbon Nanotube in Polymer Matrix Based on Developing an Equivalent Long Fiber. *Mech. Res. Commun.* **2010**, *37*, 235–240.

44. Buffa, F. Abraham, G. A. Grady, B. P. Resasco, D. Effect of Nanotube Functionalization on the Properties of Single-walled Carbon Nanotube/Polyurethane Composites. *J. Polym. Sci. Part B. Polym. Phys.* **2007**, *45*, 490–501.

45. Kalamkarov, A. L.; Georgiades, A. V.; Rokkam, S. K.; Veedu, V. P.; Ghasemi-Nejhad, M. N. Analytical and Numerical Techniques to Predict Carbon Nanotubes Properties. *Int. J. Solids Struct.* **2006**, *43*, 6832–6854.

46. ANSYS Inc. *Theory Manual*; SAS IP Inc., 2009.

47. Shokrieh, M. M. Rafiee, R. Investigation of Nanotube Length Effect on the Reinforcement Efficiency in Carbon Nanotube Based Composites. *Compos. Struct.* **2010**, *92*, 647–652.

48. Daniel, I. M.; Ishai, O. *Engineering Mechanics of Composite Materials*, 2nd ed.; Oxford University Press; New York, 2006.

49. Kelly, A.; Tyson, W. R. Tensile Properties of Fibre-reinforced Metals: Copper/Tungsten and Copper/Molybdenum. *J. Mech. Phys. Solids* **1965**, *13*, 329.

50. Hsueh, C. H. Interfacial Debonding and Fiber Pull-out Stresses of Fiber-reinforced Compos. *Mater. Sci. Eng. A* **1990**, *123*, 1–11.

CHAPTER 5

ENVIRONMENTAL ENGINEERING APPLICATIONS OF CARBON NANOTUBES: A CRITICAL OVERVIEW AND A VISION FOR THE FUTURE

SUKANCHAN PALIT*

Department of Chemical Engineering, University of Petroleum and Energy Studies, Post Office Bidholi via Prem Nagar, Dehradun 248007, Uttarakhand, India

E-mail: sukanchan68@gmail.com; sukanchan92@gmail.com

ABSTRACT

Human civilization and human scientific endeavor are today moving from one scientific paradigm toward another. The world of environmental engineering is witnessing one definite challenge over another. Technology and engineering science of nanotechnology are in the path of newer scientific regeneration. Vision and the challenge of environmental engineering applications of carbon nanotubes is extremely challenging today. Carbon nanotubes, nanomaterials, and engineered nanomaterials are in the path of newer scientific rejuvenation today. In this chapter, the author poignantly depicts the vast areas of applications of carbon nanotubes in environmental protection. Industrial wastewater treatment, drinking water treatment, and the futuristic vision of environmental sustainability will all lead a long and visionary way in the true realization of carbon nanotubes applications today. Nanotechnology is an emerging area of science that covers a wide range of technologies which are present in the nanoscale. Technological motivation and scientific profundity are the hallmark of scientific research pursuit in nanotechnology. Environmental protection today stands in the midst of vision and scientific introspection. Climate change, loss of ecological biodiversity, and the frequent environmental disasters have urged scientists throughout the world to move toward newer scientific innovations and the era of scientific emancipation in nanotechnology. Technological innovations are few in nanotechnology but the vision is bright and groundbreaking. This chapter gives a strong message toward the human scientific success, the visionary ideals and the futuristic vision in the path toward carbon nanotubes applications to environmental protection. Human scientific progress and the vast scientific vision are the torchbearers toward a newer beginning in the world of nanotechnology.

5.1 INTRODUCTION

The world of challenges in nanotechnology is groundbreaking and crossing visionary scientific frontiers. Environmental protection today stands in the midst of deep scientific comprehension and vision today. Climate change is a major challenge to human civilization today. In this chapter, the author rigorously points toward the vast success of scientific endeavor and the futuristic vision in the nanotechnology applications to human society. Human mankind and human scientific endeavor today stands in the midst of deep scientific introspection and deep disaster as environmental engineering science moves toward a newer scientific regeneration. Global climate changes, loss of ecological biodiversity, and frequent environmental disasters have urged the scientific domain to plunge into newer scientific innovations and newer scientific vision. This chapter gives a wider glimpse on the vast scientific potential, the scientific success, and the deep scientific profundity in tackling global environmental issues and for a greater scientific emancipation of nanotechnology. Nanoengineering and nanovision are the technological vision of today. Today, nanotechnology has diverse applications in all branches of science and engineering. Chemical process engineering, petroleum engineering science, and environmental engineering today are in the path of immense scientific rejuvenation and deep scientific crisis. The state of human planet's environment is disastrous and needs to readdressed and reenvisioned with the passage of scientific history and visionary timeframe. Mankind's immense scientific prowess and scientific divination, the vast technological challenges and the human needs will lead a long and visionary way in the true realization of environmental sustainability and environmental engineering science. Sustainability science as defined by Dr. Gro Harlem Brundtland, former Prime Minister of Norway needs to be redefined and reenvisioned as global concerns for environmental protection rises and surpasses visionary scientific frontiers.

5.2   WHAT DO YOU MEAN BY CARBON NANOTUBES?

Carbon nanotubes (CNTs) and other nanomaterials are today revolutionizing the scientific landscape. Scientific vision, deep scientific cognizance, and scientific forbearance are the necessities of human scientific research pursuit in environmental engineering science today. Water purification, industrial wastewater treatment, and drinking water treatment are the utmost need of the hour and are the visionary fields of scientific research pursuit. Global concerns for environmental protection are changing the veritable scientific firmament. Human civilization and human scientific endeavor are today in the midst of deep introspection and vast scientific discernment. Frequent environmental disasters, loss of ecological biodiversity, and the alarming issue of climate change are veritably in the path of scientific regeneration and immense scientific determination. CNT is tube-shaped, made of carbon having a diameter measuring on the nanometer scale. A nanometer is one-billionth of a meter, or about 10,000 times smaller than hair. CNTs are extremely unique because the bonding between atoms is extremely strong and the tubes can have extreme aspect ratios. Human scientific vision in the domain of CNTs and other engineered nanomaterials are vast and versatile. The entire chapter opens up new windows of scientific instinct and deep scientific introspection in the field of CNTs and environmental engineering.

5.3   ENVIRONMENTAL SUSTAINABILITY AND THE VISION FOR THE FUTURE

Environmental and energy sustainability are the utmost need of human progress today. Sustainable development and infrastructural development today are the hallmarks of human civilization and human scientific firmament today. The marvels and masterpieces of science and engineering are the techniques of environmental protection. Environmental engineering science and environmental sustainability are the two opposite sides of the visionary coin. Scientific regeneration and scientific rejuvenation in chemical process engineering and environmental engineering are the immediate necessities of human civilization today. Mankind's immense scientific vision, the need of human civilization, and the ever-growing global concerns for environmental protection will all lead a long and visionary way in the true emancipation of both environmental engineering science and nanotechnology. Both environmental and energy sustainability are the utmost need of the hour in the visionary path toward scientific regeneration. Water science and technology and self-sufficiency in provision of potable water encompass environmental sustainability. The vision for the future in application of advanced oxidation processes and novel separation processes is wide and bright. Membrane separation processes such as nanofiltration are veritably crossing visionary scientific boundaries.

5.4   ENVIRONMENTAL PROTECTION AND THE MARCH OF SCIENCE AND ENGINEERING

Engineering science and technology are today moving forward at a rapid pace surpassing one visionary paradigm toward another. Science of nanotechnology and CNTs are today experiencing immense scientific and technological rejuvenation. The world today stands in the midst of deep scientific vision and unanswered technological issues. Water science and technology as well as water purification are the burning issues of mankind today. Industrial wastewater treatments as well as drinking water paradigm are changing the face of human scientific endeavor today. Scientific profundity, scientific forbearance, and deep scientific cognizance are the hallmark of scientific advancement today. Human scientific progress and human technological validation today needs to be envisioned with respect to environmental engineering science and the holistic world of environmental protection. March of science and engineering today stands in the midst of deep scientific cognizance and scientific comprehension. Mankind's immense scientific prowess, the technological divination, and scientific revelation of research pursuit will all lead a long and visionary way in the true realization and true emancipation of nanoscience and nanoengineering. Today the world of environmental sustainability and the holistic world of sustainable development are in the midst of crisis and scientific introspection. Environmental protection and the world of industrial wastewater treatment need to be more holistic and groundbreaking. Arsenic and heavy metal groundwater contamination are devastating the scientific firmament today. In developing and developed nations throughout the world, arsenic groundwater remediation is the utmost need of the hour. In the similar vein, heavy metal groundwater remediation needs to be reenvisioned and reenvisaged with the passage of deep scientific history and the visionary timeframe. Technology and engineering science has a few answers toward the immense scientific stature of groundwater remediation.

5.5 SIGNIFICANT SCIENTIFIC ENDEAVOR IN THE FIELD OF ENVIRONMENTAL PROTECTION

Today the science of environmental protection, industrial pollution, and industrial wastewater treatment stands in the midst of deep scientific discernment and vast scientific cognizance. With the progress of human civilization, the challenges of environmental protection and environmental engineering science need to be reenvisioned and deeply comprehended. Nanovision and nanoengineering are today ushering in a new era in scientific firmament. Environmental protection and environmental engineering science are the utmost need of the hour for human progress. Today environmental engineering science and the world of water purification are two opposite sides of the visionary coin. Mankind's immense scientific girth, the march of science and technology, and the futuristic vision of nanotechnology will all lead a long and visionary way in the true emancipation and the true realization of environmental engineering science as well as sustainable development. Industrial wastewater treatment and drinking water treatment should be the hallmark of human scientific endeavor and human progress. The vast scientific success, the deep scientific potential, and the true emancipation of nanotechnology are the pallbearers toward a newer era in the field of both energy and environmental sustainability.

5.6 VISIONARY SCIENTIFIC ENDEAVOR IN THE FIELD OF APPLICATION OF CNTs IN ENVIRONMENTAL PROTECTION

Science and technology are today in the path of newer scientific instinct and newer scientific intuition. Environmental engineering science today stands in the midst of immense fortitude and scientific forbearance. CNTs and engineered nanomaterials are the smart materials of tomorrow. Scientific research pursuit and scientific revelation in the field of CNTs applications in environmental protection are enumerated in details in this section of this chapter.

Ren et al.[1] deeply and poignantly pondered on the subject of CNTs as adsorbents in environmental pollution management. Technological vision, scientific motivation, and the vast scientific regeneration in the field of CNTs applications in environmental protection are deeply discussed in minute details. CNTs and engineered nanomaterials are the smart materials of the present generation of science and technology.[1] Environmental pollution management today stands in the midst of deep scientific vision, scientific acuity, and scientific forbearance. Loss of ecological balance has led human civilization toward greater realization and true emancipation of environmental engineering tools. CNTs have aroused widespread worldwide attention as a new type of adsorbents due to their outstanding ability for the removal of various inorganic and organic pollutants, and radionuclides from large volumes of industrial wastewater.[1] This chapter summarizes the properties of CNTs and their properties related to the adsorption of various organic and inorganic pollutants from large volumes of aqueous solutions.[1] Their application of as adsorbents for the preconcentration and immobilization of all kinds of recalcitrant pollutants from gas streams and large volumes of aqueous solutions are summarized and the future research trends are enumerated in details. Human scientific endeavor today is at its helm as scientific destiny as nanotechnology gears forward. CNTs, a new member of the carbon family were first discovered by Iijima in 1991.[1] Technology and engineering science advanced at a rapid pace after that groundbreaking discovery. Since then, CNTs had been the focus of considerable research and immense scientific attention because of their unique physicochemical properties. The first observed CNTs were multiwalled carbon nanotubes (MWCNTs), consisting of up to several tens of graphitic shells with adjacent shell separation of approximately 0.34 nm, diameters of approximately 1 nm and large length/diameter ratio.[1] Scientific research prowess, the vast scientific vision and the basic needs of human society will all lead a long and visionary way in the true realization of CNTs and other engineered nanomaterials. MWCNTs can be considered as elongated fullerenes.[1] CNTs have been termed as "visionary materials of the 21st century" due to their unique properties such as functional, mechanical, thermal, electrical, and optoelectronic properties which depend on atomic arrangement, the diameter and the length of the tubes, and the morphology, or nanostructure.[1] Because of the current and increasing investment and potential widespread use, CNTs spread quickly in the environment. Several widespread studies suggest that they are toxic to human beings and other organisms, and their presence in the environment affects the physicochemical behavior of common environmental pollutants. Human civilization and human scientific research pursuit today stands in the midst of deep scientific vision and deep scientific fortitude. The status of nano-technology research globally is surpassing vast and versatile scientific boundaries.[1] This paper reviews the current state of applications of CNTs including MWCNTs and SWCNTs in the removal of organic and inorganic pollutants from gas and large volumes of aqueous solutions.[1] The vision of science and the technological motivation are the veritable forerunners toward a greater scientific understanding of nanotechnology and nanoengi-neering today. This chapter is a sublime and visionary effort toward greater emancipation of CNTs applications in environmental engineering science.[1] The status of research work and research forays in environmental protec-tion is in a state of deep scientific introspection. The author in this chapter deeply comprehends the wide vision, the vast scientific prowess, and the technological motivation behind CNTs application in human society.[1]

Tan et al.[2] discussed with insightful lucidity, energy and environmental applications of CNTs. CNTs applications and deep emancipation are the hallmarks of civilization's scientific prowess today.[2] Human civilization's immense environmental engineering scientific prowess, the status of environment, and the futuristic vision of environmental engineering tech-niques will lead a long and visionary way in the true realization and true emancipation of nanotechnology and nanoengineering today. Energy and environment are major global problems inducing environmental protec-tion problems.[2] Mankind's immense stature and the status of civilization's research and development initiatives are the pallbearers toward a newer era of science and engineering. Energy generation from conventional fossil fuels has been seriously identified as the main culprit of environmental quality degradation and the veritable issue of environmental pollution.[2] CNTs and their hybrid nanocomposites have received worldwide attention for their potential applications in various fields due to their unique struc-tural, electronic, and mechanical properties.[2] In this chapter, the author deeply discussed the application of nanotubes (1) in energy conversion and storage such as in solar cells, fuel cells, hydrogen storage, lithium batteries, and electrochemical supercapacitors, (2) in environmental moni-toring and wastewater treatment, and (3) green nanocomposite design.

This chapter gives a wide overview of the advantages imparted by CNTs in electrochemical devices of energy applications and green nanocomposites, as well as nanosensor and adsorbent for environmental protection.[2]

Kunduru et al.[3] deeply discussed with lucid and cogent insight nanotechnology for water purification and application of nanotechnology methods in wastewater treatment. Water is the most important and pivotal asset of human civilization and pure drinking water is a basic human necessity. Technological vision and deep scientific motivation is the cornerstone of this well-researched chapter. Water contaminants may be organic, inorganic, and biological.[3] Some contaminants are toxic and vehemently carcinogenic. Aromatic hydrocarbons such as benzene, toluene, ethyl benzene, and xylenes, labeled as BTEX are the most predominant components in produced wastewater among other volatile components.[3] Human scientific prowess and scientific endurance today stands in the midst of vision and deep comprehension. The authors give a wide overview of different nanomaterials in water and wastewater treatment. The challenge and vision of nanotechnology is scientifically inspiring as science and technology initiatives moves ahead. The salient and sublime features of this chapter are the application of nanotechnology methods in industrial wastewater treatment and the discussion on the major limitations associated with conventional water purification methods. Human scientific regeneration, the vast scientific far-sightedness, and the futuristic vision of nanotechnology are the torchbearers of this well-researched chapter.[3]

Suthar et al.[4] discussed and delineated with vast scientific foresight nanotechnology for drinking water purification. Water is an essential but often overlooked nutrient in the human diet and the human body. In fact, the average adult human requires the consumption of approximately 2 L of water daily to maintain the body's respiratory balance. Technology and engineering science today stands in the midst of deep scientific introspection. Potential water scarcity and unending water crisis are being recognized as threats to human activity.[4] The challenge and the vision of human scientific endeavor needs to be reenvisioned and reenvisaged as regards application of environmental engineering science and environmental sustainability. Water resources continue to diminish due to overuse, pollution, and immense waste. Here comes the importance of nanotechnology applications and the visionary domain of environmental sustainability. This chapter highlights the recent developments of nanoscale materials and processes for treatment of subsurface water, ground water that are plagued by contaminants, such as toxic metals, organic and inorganic compounds, heavy metals, bacteria, and viruses. Human scientific regeneration and human scientific endeavor are today witnessing immense challenges as science and technology surges forward.[4] The Environment Protection Agency (EPA) categorizes drinking water contaminants into the following categories: organic, inorganic, disinfectants, disinfectant by-products, microorganisms, and radionuclides. The entire domain of environmental chemistry today stands in the midst of deep scientific vision and introspection. This chapter deeply unravels the scientific success, the vast scientific potential, and the visionary challenges in nanotechnology applications in water purification. The authors in this chapter discussed adsorption, application of carbon nanomaterials, CNTs, metal-based nano-adsorbents, natural polymer nanocomposites, synthesized polymer nanocomposites, and the wide domain of membranes.[4] Scientific vision, deep scientific discernment, and the futuristic vision of nanotechnology will lead a long and visionary way in the true emancipation of environmental engineering science today.[4]

Figoli et al.[5] deeply discussed with scientific zeal and determination application of nanotechnology in drinking water purification. Safe, reliable, and scientifically sound water supply is an important challenge for the progress of human civilization. Demand for clean water resources is veritably increasing owing to the rapid industrialization of developing countries, population growth, and long-term droughts in developing countries. The authors in this chapter vastly pondered upon antimicrobial nanotechnology in water disinfection. The vision of engineering science, the technological profundity, and the success of nanotechnology will all lead a long and effective way in the true emancipation of environmental engineering and the science of water purification today. Water technology today stands in the midst of deep scientific introspection and academic rigor. Science and technology of water purification today stands in the midst of deep scientific revelation and unending scientific rigor.[5] This is a sublime and subtle portrayal of the immense scientific intricacies behind the science of water purification and a deep understanding of the vast and varied nanotechnology applications in industrial wastewater treatment and the drinking water paradigm. Human scientific research pursuit in drinking water purification and water pollution control in developing countries needs to be vehemently addressed and reenvisioned with the passage of scientific history and the visionary timeframe. In this chapter the author also delves deep into the latent domain of advanced oxidation processes with nanostructured photocatalysts.[5] The other salient features of this well-researched chapter are the visionary domain of membranes for water purification and the nanotechnology applications. The areas of membrane science delved upon are nanostructured photocatalytic membranes, electrospun nanofiber membranes, CNT membranes, graphene membranes, zeolite membranes, and the vast area of nanocomposite membranes.[5] Nanoadsorbents for water purification are the other scientific landscape of this well-researched chapter. Human scientific emancipation and the deep need of human society will today lead a long way in the true realization of nanotechnology and water purification.[5]

5.7   VISIONARY SCIENTIFIC RESEARCH PURSUIT IN THE FIELD OF INDUSTRIAL WASTEWATER TREATMENT

Industrial wastewater treatment, drinking water treatment, and the wide world of water purification today stand in the midst of deep scientific comprehension and definite vision. Environmental engineering science, chemical process engineering, and nanotechnology are today linked with each other with immense scientific vision and forbearance. The challenges and vision of human society has urged engineers and technologists to move toward more scientific innovations and more scientific intuition. The huge domains of water purification are facing immense scientific and technical challenges. Global water crisis is at a state of immense distress and deep comprehension. Industrial wastewater treatment today stands in the crucial juncture of vision and deep scientific adjudication. Today the environmental engineering world needs to be reframed and restructured as regards research and development initiatives. Traditional and nontraditional environmental engineering tools are today changing the scientific firmament of scientific vision and scientific forbearance. Technology and engineering science of nanotechnology applications in environmental protection are ushering in a new era of scientific regeneration and rejuvenation. In this well-researched chapter, the author deeply unravels the challenges and the vision behind today's application of nanotechnology in the furtherance of science and engineering. Human society at this critical juncture of environmental engineering science is poised toward newer vision and newer innovation.

Sarkar et al.[6] deeply discussed with lucid insight nanotechnology-based membrane separation process for drinking water purification. The scarcity of drinking water has become a grave concern globally due to rapid proliferation of the global population and consequently in last century, it has been estimated that the rate of water consumption has been increased up to seven times to its initial rate. Scientific vision and deep scientific fortitude are the necessities of scientific research pursuit in water purification today.[6] According to a survey of the World Water Council, about 4 million people will suffer water scarcity by 2030 due to rapid increase of water use in domestic purpose, agriculture, industry, energy sector, and so on. Human civilization and the research forays in environmental protection today veritably stand in the midst of deep turmoil and introspection. The authors poignantly depicted a brief history of membranes, organic membranes, thin-film composite membranes, nanoparticle-based membranes, inorganic–organic membranes, and biologically inspired membranes.[6] Scientific motivation and technological profundity are the deep hallmarks of this chapter. The other cornerstones of this chapter are the visionary pathways for commercialization of nanomaterial-based membranes.[6]

Palit[7] discussed with immense foresight application of nanotechnology, nanofiltration, and drinking and wastewater treatment. Scientific fortitude today is in the path of newer regeneration and deep restructuring. The world of environmental engineering science is veritably moving steadily and steadfastly through visionary challenges and insurmountable barriers.[7] Global water crisis has challenged and changed the scientific landscape and deep scientific firmament of environmental engineering science. In this chapter, the author with much vision and academic rigor unfolds the technological marvels behind nanotechnology and nanofiltration.[7] The author in this chapter targets global drinking water crisis, difficulties, and vast success of novel separation processes. The intricacies of science and engineering of membrane science and technology today are slowly unfolding. Some of the salient and sublime features of this chapter are (1) global initiative in drinking water provision to common mass, (2) environmental sustainability and human scientific progress, (3) application of nanotechnology in drinking water treatment, (4) arsenic groundwater remediation in developing world, and (5) effective novel separation processes in drinking water treatment.[7]

Research and development initiatives in water purification and industrial wastewater treatment are the hallmarks of environmental engineering today. In this chapter the author lucidly comprehended the vast scientific genre, the scientific paradigm, and the scientific vision behind nanotechnology applications in water purification. Today science is a huge colossus with a definite and purposeful vision of its own. This chapter is a veritable eye-opener toward the vision of environmental engineering and nanotechnology.

5.8   NANOTECHNOLOGY AND ENVIRONMENTAL PROTECTION

Nanotechnology, environmental engineering science, and the holistic world of environmental protection are the utmost need of human society today. Human scientific progress, mankind's immense scientific prowess, and the futuristic vision of technology and engineering science will lead a long and visionary way in the true emancipation and true realization of environmental protection. The true validation of science and engineering are the two opposite sides of the visionary coin. Degradation of recalcitrant inorganic and organic compounds is of immense concern to the human society and the human scientific progress in particular. Membrane science particularly nanofiltration today assumes immense and primordial importance with the progress of science of environmental protection. The challenge and the vision of traditional and nontraditional environmental engineering techniques are immense, ever-growing, and groundbreaking. Novel separation processes and advanced oxidation processes are changing the face of scientific endeavor today. The world and the human civilization today are in the midst of difficulties and immense environmental crisis. Technology and engineering science needs to be reenvisioned and reenvisaged as human civilization progresses. Today nanotechnology is linked with diverse branches in engineering science such as chemical process engineering, environmental engineering science, and petroleum engineering. Technological motivation, scientific acuity, and scientific validation are today the hallmarks of a newer scientific regeneration and a newer visionary eon in engineering science. Nanofiltration and membrane science are veritably linked with environmental engineering science with immense vision and scientific vigor. Today human civilization stands in the midst of deep scientific crisis and unending quagmire. Technology has few answers to the intricacies of nanotechnology, nanomaterials applications, and CNTs applications. Here arises the immense importance of scientific and technological validation. Environmental protection and environmental engineering science are the need of the hour in the true emancipation of engineering and science today.

5.9   GLOBAL CONCERNS FOR PROVISION OF PURE DRINKING WATER

Provision for pure drinking water is an important parameter toward human scientific progress today. Technological and scientific barriers are the challenges and vision of human society in present day human civilization. The scientific success, the scientific aura, and the scientific ardor need to be reenvisioned with the progress of science of sustainability. Provision of clean drinking water is an imperative to the human scientific progress today. The ardor of human society, the vast technological vision, and the world of scientific validation will go a true and visionary way in the true realization of environmental and energy sustainability. Water purification is a global issue as science and technology moves forward. Environmental and energy sustainability are the success of human scientific endeavor and human civilization today. Water purification, the need for pure drinking water, and the futuristic vision of industrial wastewater treatment will lead a definite path toward environmental sustainability.

One of the difficult problems afflicting people throughout the world is inadequate access to clean and pure water and sanitation. Shannon et al.[8] deeply comprehended and lucidly delineated the science and technology for water purification in the coming decades. Burning problems with water are expected to grow worse in the coming decades, with water scarcity occurring globally and thus, the whole domain of water purification needs to be reenvisioned and readdressed with the passage of human history and time. Deeply addressing these problems calls out for a tremendous amount of research pursuit to be conducted to identify robust methods of purifying water at lower cost and with less energy while at the same time and same situation minimizing the use of chemicals and the adverse impact on the environment. Human scientific research pursuit today stands in the midst of vision and vast academic rigor. Water purification in the similar manner needs to be deeply comprehended and vastly addressed as science and engineering surges ahead. In this chapter (Shannon et al., 2008),[8] the authors lucidly highlights some of the science and technology being developed to improve the disinfection and decontamination of water, as well as the efforts to increase water supplies through the safe reuse of wastewater and effective desalination of sea and brackish water. The immense problems worldwide associated with the lack of clean and fresh water are well known: 1.2 billion people lack access to safe drinking water, 2.6 billion have little or no sanitation, millions of people die annually—3900 children a day—from diseases transmitted through unsafe water or human excreta. Technology and engineering science today are the cornerstones of human civilization and human scientific endeavor. Water hiatus still today stands a major impediment to human scientific progress. Technology has a few answers to the world of challenges and scientific barriers in water purification science today. Water also strongly affects energy and food production, industrial output, and the quality of our environment, affecting the developing and developed economies. Science and technology today are in the path of deep regeneration and vast scientific rejuvenation. Many freshwater aquifers are being deeply contaminated and overdrawn in populous regions—some irreversibly—or suffer saltwater intrusion along coastal regions. With agriculture, livestock, and energy veritably consuming more than 80% of all water for human use, demand for fresh water is expected to increase with immense population growth, further stressing traditional resources. Shannon et al.[8] deeply and lucidly enumerated the domains of disinfection, decontamination, reuse and reclamation, and desalination. Desalination is a major research thrust area in present-day human civilization. Water-stressed countries throughout the world are veritably dependent on new and innovative environmental engineering tools which include traditional and nontraditional techniques. Human scientific progress, the futuristic vision of environmental engineering science and the overarching goals of human life will surely lead a long and visionary way in the true emancipation of science and technology today. The author (Shannon et al.),[8] mostly concentrated in this chapter on the vast scientific potential of environmental engineering and water purification tools in a clear understanding and vision toward advancement of science and technology.

Hashim et al.[9] deeply discussed with lucid and cogent insight remediation technologies for heavy metal contaminated groundwater. Green engineering and green nanotechnology are the forerunners toward a newer visionary era in the field of science and engineering today. Arsenic and heavy metal groundwater contamination are the burning issues of environmental engineering paradigm today. The contamination of groundwater by heavy metal, originating either from natural soil sources or from anthropogenic sources is a matter of grave concern to the public health and the human society.[9] Remediation of contaminated groundwater is of utmost priority since billions of people around the world use it for drinking water purpose. Water science and technology are the hallmarks of scientific endeavor in environmental engineering today. Human scientific progress today stands in the midst of vision and forbearance. Scientific endeavor in today's world needs to be reenvisioned as global water crisis devastates the wide scientific firmament.[9] Human scientific intelligence, scientific prowess, and the world of technological validation will veritably lead a long way in the true realization of environmental protection and industrial pollution control. In this chapter, the author deeply and lucidly ponders upon 35 approaches of groundwater treatment and classified them in three broad categories such as chemical, biochemical, and physicochemical treatment processes.[9] Validation of science is the necessity of scientific research pursuit today. Human vision as regards energy and environmental sustainability needs to be redrafted and reenvisaged with the progress of scientific history and time. Selection of a suitable technology for contamination remediation at a particular site is one of the challenging jobs due to extremely complex soil chemistry and aquifer characteristics. The intricacies in groundwater remediation science go beyond scientific imagination. Technology and engineering science has few answers to the vast scientific fabric of groundwater remediation. Keeping the vast sustainability issues in mind and the grave concern of environmental ethics, the technologies encompassing natural chemistry, bioremediation, and biosorption are recommended to be adopted in relevant areas. The areas of applied science such as applied chemistry thus need to be vehemently addressed. In this chapter, the authors delved deep into sources, chemical property, and speciation of heavy metals in groundwater. The other areas the author deeply discussed are technologies for treatment of heavy metal contaminated groundwater, chemical treatment technologies, in situ treatment by using reductants, reduction by using iron-based techniques, soil washing, in situ soil flushing, in situ chelate flushing, and biological, biochemical, and biosorptive treatment technologies. Human scientific determination and fortitude need to be restructured and revamped as scientific and academic rigor as groundwater remediation surges forward.[9]

Chakraborti et al.[10] lucidly delineated the vast and visionary domain of India's groundwater quality management. India is a developing country and today poised for a visionary economic growth. The nation today has immense water quality management issues. The authors in this chapter with vision and scientific rigor enumerates on the present groundwater quality in India. Today, globally groundwater remediation stands as a major scientific achievement in the scientific firmament of environmental engineering science.[10] Technology and engineering science stands in the midst of enigma and scientific vigor. The world of challenges in groundwater quality assessment and management in a developing country like India need to be vehemently addressed and reenvisaged as science and engineering surges forward. The Government of India's (GOI) Environmental Hygiene Committee, constituted shortly after India's independence from the British rule to address an immense challenge which is estimated in a 1949 Report that cholera, dysentery, and diarrhea were alone responsible for over 40,000 deaths in India from 1940 to 1950.[10] Human scientific destiny and human scientific provenance in the global water and energy scenario are slowly gearing up toward immense challenges. The committee recommended, in addition to immediate measures, the provision of potable water to at least 90% of India's population within 40 years.[10] Watershed management moved toward new and visionary scientific avenues. Nanotechnology is one of the many answers toward successful environmental sustainability today. To surmount the immense problems greatly impeding economic growth and overall development, the national government, in association with international agencies, began a transition to the utilization of the country's abundant groundwater resources. Thus, science ushered in a new era in the field of environmental engineering.[10]

5.10   WATER PURIFICATION, NANOTECHNOLOGY, AND ENVIRONMENTAL SUSTAINABILITY

Nanoscience and nanotechnology are in the path of new scientific regeneration and a newer visionary era. In the present-day human civilization, water purification and nanotechnology are the two opposite sides of the visionary and pathbreaking coin. Human civilization and human scientific endeavor in sustainability science needs to be reenvisioned and reenvisaged as human scientific progress surges forward. Environmental sustainability and water purification are today two opposite sides of the visionary coin. In a detailed United Nations Organization Report,[11] the authors deeply discussed with immense foresight the utmost need of water and energy sustainability globally. Sustainable development and sustainability are two visionary avenues of scientific introspection and vision. Reducing poverty, ensuring economic growth, and building up a more inclusive society are the outstanding collective achievements of human civilization today. Global water crisis and global water hiatus needs to be reenvisioned and vehemently readdressed as science and engineering gears forward. This chapter is a veritable eye-opener toward the global water crisis and the advancement of science and technology. Success in economic growth and economic advancements requires harnessing the immense potential of ecosystems to satisfy the demands of energy and water which are essential for human progress as well as for the functioning of the many production and consumption processes where water and energy are the pivotal components. Mankind's immense scientific prowess, man's immense scientific zeal, and the futuristic vision of water science and technology will lead a long and visionary way in the true emancipation and true emancipation of environmental sustainability. Today, nanotechnology is in the forefront of human scientific endeavor. Nanotechnology and sustainability are the two opposite sides of the visionary coin. This chapter poignantly depicts the human success, the management of the environmental impacts of water and energy, the environmental effects of water and energy, policy instruments, risk management, and the futuristic vision of research and development initiatives.[11] Human civilization's immense scientific grit and determination, the vast scientific adjudication, and the technological validation are the torchbearers toward a newer visionary era in the field of water and energy. Risk management is another hallmark of this well-researched chapter.

U. S. Environmental Protection Agency Report[12] deeply discussed with lucid foresight energy trends in selected manufacturing sectors. The base case scenario was petroleum refining. The petroleum refining industry includes establishments engaged in refining crude petroleum into refined petroleum products through multiple distinct processes including distillation, hydrotreating, alkylation, and reforming.[12] The upshot of this chapter is the greater scientific emancipation in the field of energy sustainability. Energy self-sufficiency is the crux of human scientific progress today. This report with deep scientific vision delineates the energy scenario as respects petroleum refining.[12]

World Energy Council Report[13] discussed and elaborated on the Energy Sustainability Index. This report provides country-level details on the results of the 2013 Energy Sustainability Index prepared by the World Energy Council. Scientific vision, deep scientific fortitude, and the vast scientific imagination are the pallbearers toward a newer era in energy and environmental sustainability today. Energy sustainability dimensions include energy security, energy equity, and environmental sustainability. The visionary words of Dr. Gro Harlem Brundtland, former Prime Minister of Norway on sustainability are the hallmarks of human scientific progress and gears toward true emancipation of energy and environmental sustainability.[13] The success of human civilization today depends wholly upon true realization of energy and environmental sustainability. The science of sustainability needs to be redefined and readdressed. Resolving global water issues is a major scientific achievement in present-day human civilization. The futuristic vision of energy sustainability, the scientific needs of human society, and the global concerns for environmental protection will lead a long and visionary way in the effective realization of global energy security. Human scientific progress today stands between deep introspection and vision.[13] Technology needs to be redefined and readdressed as regards proper realization of sustainable development. This report also reflects the results of the 2013 Energy Sustainability Index prepared by World Energy Council.[13] The Index evaluates how well countries throughout the world balance the three often conflicting goals of energy sustainability—energy security, energy equity, and environmental sustainability—what the World Energy Council defines the "energy trilemma." Human scientific vision as regards energy sustainability today is in the path of newer scientific and technological regeneration. This World Energy Council Report also redefines public stakeholder recommendations. The salient and sublime features of the report go beyond cross-boundary research and opens up new windows of innovation and instinct in decades to come.[13]

The Third World Academy of Sciences Report[14] poignantly depicted the sustainability science landscape in the developing world. Energy sustainability is the utmost need of the hour in developing world today.[14] Mankind's immense scientific prowess, the needs of human scientific progress, and the challenges of energy security will veritably lead a long way in the visionary road to successful sustainability. This report delineates the concept of global energy trends, the transition to commercial forms of energy, the needs for improving energy efficiency, decarbonization and diversification, the sustainable energy challenge from the developing country perspective, and the veritable concerns of policies to promote the proper implementation of sustainable energy technologies.[14]

Human civilization's immense scientific grit and determination, the challenges and vision of environmental engineering and energy engineering, and the scientific urge to excel are the pivotal parameters toward human development today. Throughout this chapter, the author rigorously points out toward the nanotechnology applications of science in present-day human civilization. The technological challenges and the vast scientific motivation need to be redefined as civilization surges forward. Environmental protection, at this crucial juncture of scientific history and visionary timeframe, is facing immense scientific intricacies and challenges. This chapter is a veritable eye-opener toward mankind's deep introspection into nanotechnology and environmental engineering science.

5.11 RESEARCH ENDEAVOR, SCIENTIFIC VALIDATION, AND MEMBRANE SCIENCE

Research endeavor, scientific validation, and membrane science are the utmost need of human society as human civilization moves forward. Novel separation processes such as membrane science and nanofiltration are the visionary avenues of scientific research pursuit today. Research validation and scientific and technological motivation are the successful pathways toward true emancipation of environmental engineering science in present-day human civilization. Membrane science and nanotechnology are today veritable parameters toward the advancement of environmental engineering science. The areas of membrane science in today's perspective need to be reenvisioned and reenvisaged with the passage of scientific history and time. It is reported that 96.5% of the earth's water is located in seas and oceans, 1.7% in the ice caps, 0.8% is considered to be fresh water with the rest being brackish water. Since water shortage has been a serious problem for human society and human progress and humans have been searching for the solution for a long time, desalination, turning salty water into fresh water, is not a new scientific endeavor. Today the worldwide shortage of drinking water is a serious worldwide concern due to population proliferation and the increased demand for potable drinking water that exceeds readily available water resources. Science and technology of water purification is highly challenged today. Scientific validation and technological profundity is the necessity of human progress today. Reverse osmosis, nanofiltration, and electrodialysis are the typical membrane processes available for desalination. Reverse osmosis and nanofiltration are called pressure-driven membrane processes since the transmembrane pressure difference is the driving force for the mass transport while for electrodialysis the electrical potential difference is the driving force for the mass transport. These are the primordial factors toward the successful realization of membrane separation phenomenon.

5.12   TRADITIONAL WASTEWATER TREATMENT TECHNOLOGIES

Traditional and nontraditional wastewater treatment technologies are challenging the scientific firmament and the vast scientific fabric of environmental engineering science today. The area of membrane science falls under traditional wastewater treatment technology. Scientific vision, deep scientific divination, and the vast technological profundity in the field of environmental engineering science will lead a long and visionary way in the true emancipation and true realization of sustainability science. The vast academic and scientific rigor in the field of membrane science today need to be redefined as science and technology surges forward toward a newer eon. Wastewater is widely recognized as one of significant and reliable water resources. Technological profundity in the field of membrane science is today ushering in a new era in human scientific progress. The vast area of wastewater reclamation management has tremendous scope of balancing water availability and water demand, at low cost and with lower environmental impacts. By taking into consideration the need of wastewater reclamation processes, type of membrane processes and commercial application of membrane technology needs to be reenvisioned and readdressed. A wide variety of membrane processes can be categorized according to driving force, membrane material, membrane type and configuration, removal capabilities and mechanism, and membrane fouling and cleaning. For example, pressure-driven membrane processes include microfiltration, nanofiltration, ultrafiltration, and reverse osmosis. Among these techniques, membrane bioreactor and reverse osmosis processes are veritably well suited for application of wastewater reclamation process. The first historic discovery of Loeb–Sourirajan asymmetric reverse osmosis membrane enabled seawater desalination at a larger and pathbreaking scale. Today visionary scientific frontiers need to be surpassed. Here comes the immense need of membrane science for the furtherance of science and engineering.

5.13 NONTRADITIONAL WASTEWATER TREATMENT TECHNIQUES

Human scientific research pursuit today stands in the midst of deep scientific vision and vast scientific forbearance. Environmental engineering science and its applications are the utmost need of the hour as science and technology surges forward. Advanced oxidation techniques are one of the nontraditional environmental engineering tools. Advanced oxidation processes (AOP) are characterized by a common chemical feature: the capability of exploiting the high reactivity of HO radicals in driving oxidation processes which are suitable for achieving the complete abatement and total mineralization of even less reactive pollutants. Integrated advanced oxidation processes are also part and parcel of nontraditional wastewater treatment techniques. In the last 20 years, a rather fast evolution of the immense research activities devoted to environmental protection has been recorded as the vast consequence of the special attention paid to environment by social, political, and legislative international authorities leading in some cases to the delivery of severe regulations and environmental restrictions.[15,16] Human society and human scientific research pursuit in environmental engineering science needs to be reenvisioned and revamped as science and engineering moves forward. The total degradation of toxic pollutants as also that of the simple biologically recalcitrant compounds must be therefore demanded to other nonbiological techniques. These gamut of technologies consists mainly of conventional phase separation techniques (adsorption processes and stripping techniques) and methods which destroy the contaminants (chemical oxidation/reduction). Chemical oxidation aims at the mineralization of the contaminants to carbon dioxide, water, and inorganics or, at least, at their transformation into harmless compounds. The success of human scientific endeavor, the challenges of technological prowess and the immense scientific needs of the human society will all lead a long and visionary way in the true realization of environmental engineering tools and the successful realization of environmental sustainability. Science today is a huge colossus with a deep and wide vision of its own. In this chapter, the author rigorously points out toward the vast application of nanoscience and nanotechnology in the ever-growing and pathbreaking world of environmental engineering science.

5.14 DESALINATION, WATER TREATMENT, AND THE VISION FOR THE FUTURE

Desalination and water treatment are the hallmark of environmental engineering science and environmental protection in today's human civilization. Lack of potable drinking water is the enigmatic issues of science and engineering today. The boon of human civilization and the focal point of scientific research pursuit today are the environmental engineering science and the diverse environmental engineering tools. Shannon et al.[8] lucidly discussed the ever-growing and groundbreaking domain of desalination. The overarching goal for the future of desalination is to increase the fresh water needs via desalination of seawater and saline aquifers. These sources account for 97.5% of all water on the earth, so capturing even a tiny fraction could have a huge impact on water scarcity. Through rapid scientific forays and deep scientific introspection, particularly in the last decade, desalination technologies can be used reliably to desalinate sea water as well as brackish waters from saline aquifers and rivers. Desalination of all types, though is often considered a capital- and energy-intensive process, and typically requires the conveyance of the water to the desalination plant, pretreatment of the intake water, disposal of the concentrate (brine), and process maintenance. The major desalination technologies currently in use are based on membrane separation via reverse osmosis and thermal distillation (multistage flash and effect distillation), with reverse osmosis accounting for over 50% of the installed capacity. Scientific research pursuit in reverse osmosis is today crossing vast and versatile scientific frontiers. Desalination is the need of many developing countries in satisfying water needs and provision of pure drinking water. In this chapter, the author deeply ponders upon the scientific success, the deep scientific profundity, and the scientific revelation in the field of desalination and water treatment.

5.15   FUTURE OF NANOTECHNOLOGY APPLICATIONS IN INDUSTRIAL WASTEWATER TREATMENT

Nanotechnology and nanoengineering are the scientific frontiers of human civilization today. Human scientific endurance and deep scientific vision are the hallmarks of scientific research pursuit in nanotechnology and engineered nanomaterials today. Science is a huge colossus with a definite vision of its own. Today nanotechnology and engineered nanomaterials are the marvels of science and engineering. Scientific endurance, deep scientific profundity, and the world of scientific challenges are the hallmarks of human civilization today. Technology has a few answers to the intricacies of nanotechnology. Here comes the immense importance of scientific efforts in nanoscience and the holistic domain of applied science. Industrial wastewater treatment and drinking water treatment are in a state of immense scientific vision and scientific provenance. Future of nanotechnology applications in water purification and industrial wastewater treatment need to target more toward vision and efficiency of these water treatment tools. AOP and novel separation processes are the need of human society today. Environmental engineering science and environmental protection are in the path of deep scientific rejuvenation and scientific vision. Frequent environmental disasters and loss of ecological biodiversity has urged scientists and engineers to move toward newer vision and newer innovations. Nanotechnology will surely change the vast scientific firmament of human progress. Future of nanotechnology applications in industrial wastewater treatment is extremely groundbreaking and gearing toward a visionary era. Technological vision, scientific profundity, and scientific divination will veritably lead a long and visionary way in the true emancipation of nanotechnology and environmental engineering science.[15–17]

5.16   FUTURE SCIENTIFIC RECOMMENDATIONS IN NANOTECHNOLOGY AND CNTs APPLICATIONS

Technology and engineering science of the vast domain of nanotechnology are surpassing visionary boundaries today. The challenge and the vision of nanotechnology and CNTs applications in environmental protection need to be restructured and reenvisioned with the passage of scientific history and time. Future recommendations of nanotechnology and CNTs applications should be more rigorous and replete with scientific and academic rigor. The sublime and salient feature of this chapter goes beyond scientific imagination and vast scientific forbearance. Human scientific grit, scientific determination, and the world of scientific revelation are the pallbearers toward a newer visionary era in both nanotechnology and environmental engineering science. Future targets should be toward newer innovation and newer techniques in traditional and nontraditional environmental engineering science. Novel separation techniques such as membrane separation processes need to be more envisioned and readdressed with the course of scientific history and time. Zero-discharge norms are the utmost need of the hour. Industrial wastewater treatment, drinking water treatment, and the entire domain of water purification are the challenge and the vision of present-day human scientific endeavor. CNTs are the next generation smart materials. This chapter goes beyond scientific imagination and opens up new vistas of scientific regeneration in decades to come. Human civilization and human scientific endeavor thus will surely usher in a new eon in environmental engineering science and nanotechnology. Human scientific destiny and scientific vision are in the path of newer rejuvenation today. In this chapter, the author deeply comprehends the scientific success, the wide scientific potential, and the major scientific advancements in the application of nanotechnology in environmental protection.[15–17]

5.17   CONCLUSION AND FUTURE PERSPECTIVES

Human civilization and human scientific endeavor today stand in the midst of deep vision and vast scientific comprehension. The challenge and the vision of scientific research pursuit in CNTs and environmental engineering science need to be restructured and reenvisaged with the course of scientific history and time. Technological vision, profundity, and deep scientific motivation are the necessities of human scientific research pursuit today. Depletion of fossil fuel resources, the loss of ecological biodiversity, and the enigmatic issue of global climate change are today urging the vast scientific domain to surge toward newer innovation and a newer scientific paradigm. In this chapter, the author rigorously points out toward the scientific imagination, the vast scientific cognizance, and the scientific profundity behind nanotechnology applications in environmental engineering science today. The author deeply reviews the vast scientific understanding and discernment in the research pursuit in CNTs. Human civilization and human research pursuit in the vast world of nanotechnology are today the torchbearers of scientific progress. Scientific and technological advancements in CNTs are the cause of immense scientific enigma today. It has diverse applications in all branches of science and engineering. This chapter focuses on the visionary success, the vast scientific fortitude, and the immense challenges behind nanomaterials application in mankind. Today the needs of the human civilization are the applications of frontiers of science. Environmental engineering, chemical process engineering, petroleum engineering, and nanotechnology are the frontier domains of science. Human scientific research pursuit in nanotechnology stands in the midst of deep scientific provenance and vast revelation. This chapter gives a glimpse of the immense scientific applications of CNTs and vistas of nanotechnology. Scientists and engineers are the masterminds of human civilization today. Technological advancements, the deep need of scientific endeavor, and the futuristic vision of nanotechnology will veritably lead a long and visionary way in the true emancipation of science and engineering in present-day human civilization. The challenge of nanotechnology and environmental engineering science are ever-growing and surpassing visionary boundaries. The author with deep and sublime effort discussed the intricacies of scientific endeavor in the field of CNTs application in environmental protection.

KEYWORDS

- water
- carbon nanotubes
- vision
- global
- nanotechnology

REFERENCES

1. Ren, X.; Chen, C.; Nagatsu, M.; Wang, X. Carbon Nanotubes as Adsorbents in Environmental Pollution Management. *Chem. Eng. J.* **2010,** *170*, 395–410.
2. Tan, C. W.; Tan, K. H.; Ong. Y. T.; Mohamed, A. R.; Zein. S. H. S.; Tan, S. H. Energy and Environmental Applications of Carbon Nanotubes. *Environ. Chem. Lett.* **2012,** *10*, 265–273.
3. Kunduru, K. R.; Nazarkovsky, M.; Farah, S.; Pawar, R. P.; Basu, A.; Domb, A. J. Nanotechnology for Water Purification: Applications of Nanotechnology Methods in Wastewater Treatment (Chapter 2). In *Water Purification;* Grumezescu A. M., Ed.; Academic Press, Elsevier: Amsterdam, Netherlands, 2017; pp 33–74.
4. Suthar, R. G.; Gao, B. Nanotechnology for Drinking Water Purification (Chapter 3). In *Water Purification;* Grumezescu A. M., Ed.; 2017; pp 75–118.
5. Figoli, A.; Dorraji, M. S. S.; Amani-Ghadim, A. R. Application of Nanotechnology in Drinking Water Purification (Chapter 4). In *Water Purification;* Grumezescu A. M., Ed.; 2017; pp 119–167.
6. Sarkar, S.; Sarkar, A.; Bhattacharjee, C. Nanotechnology Based Membrane Separation Process for Drinking Water Purification (Chapter 10). In *Water Purification;* Grumezescu A. M., Ed.; 2017; pp 355–389.
7. Palit, S. Application of Nanotechnology, Nanofiltration, and Drinking and Wastewater Treatment—a Vision for the Future (Chapter 17). In *Water Purification;* Grumezescu A. M., Ed.; 2017; pp 587–620.
8. Shannon, M. A.; Bohn, P. W.; Elimelech, M.; Georgiadis, J. G.; Marinas, B. J.; Mayes, A. M. *Science and Technology for Water Purification in the Coming Decades*. Nature, Nature Publishing Group: USA, 2008; pp 301–310.
9. Hashim, M. A.; Mukhopadhyay, S.; Sahu, J. N.; Sengupta, B. Remediation Technologies for Heavy Metal Contaminated Groundwater. *J. Environ. Manage.* **2011,** *92*, 2355–2388.
10. Chakraborti, D.; Das, B.; Murrill, M. T. Examining India's Groundwater Quality Management. *Environ. Sci. Technol.* **2011,** *45*, 27–33.
11. United Nations Report. *Water and Energy Sustainability;* 2014. www.un.org./waterforlifedecade/water_and_energy_2014 (accessed Sep 1, 2017).
12. United States Environmental Protection Agency Report. *Energy Trends in Selected Manufacturing Sectors: Opportunities and Challenges for Environmentally Preferable Energy Outcomes*; March, 2007.
13. World Energy Council Report. *2013 Energy Sustainability Index;* 2013.
14. The Academy of Sciences for the Developing World Report. *Sustainable Energy for Developing Countries*; 2008.
15. Palit, S. Nanofiltration and Ultrafiltration—the Next Generation Environmental Engineering Tool and a Vision for the Future. *Int. J. Chem. Tech. Res.* **2016,** *9* (5), 848–856.
16. Cheryan, M. *Ultrafiltration and Microfiltration Handbook*. Technomic Publishing Company Inc: USA, 1998.
17. Van der Bruggen, B.; Manttari, M.; Nystrom, M. Drawbacks of Applying Nanofiltration and How to Avoid Them: A Review. *Sep. Purif. Technol.* **2008,** *63*, 251–263.

CHAPTER 6

ADVANCES IN CARBON NANOTUBE-BASED CONDUCTING POLYMER COMPOSITES

SAJID IQBAL[1], RANGNATH RAVI[1], ANUJIT GHOSAL[2*], JAYDEEP BHATTACHARYA[2], and SHARIF AHMAD[1*]

1Materials Research Laboratory, Department of Chemistry, Jamia Millia Islamia, New Delhi 110025, India

2Plant Nanobiotechnology Laboratory, School of Biotechnology, Jawaharlal Nehru University, New Delhi 110067, India

*Corresponding author.
E-mail. anuj.ghosal@gmail.com, Sharifahmad_jmi@yahoo.co.in

ABSTRACT

The present chapter highlights the current development in the field of conducting polymer/carbon nanotube (CP/CNT) nanocomposites along with their wide range of properties. The unique electronic, chemical, and physical properties of CP and CNT have drawn the attention of scientific world. A hybrid composite of these two electrically conducting material have resulted in development of new and advanced characteristics, in turn, their application in innovative technologies. In addition to this, briefly, the future prospects of these materials have also been discussed.

6.1 INTRODUCTION

Conducting polymers (CPs) are the conjugated polymers having π-bonds on their polymeric backbone. Since, their discovery (1977), they have attracted great attention to the scientist and technologist because of their unique optical and electrical properties.[1] Among the different CPs shown in Figure 6.1, polyaniline, polypyrrole (PPy), and polythiophene (PTh) are widely used in many applications such as sensors, supercapacitors, batteries, solar cell, photocatalyst, electromagnetic shielding, anticorrosive coatings, etc.[2-7] The CPs are lightweight, flexible, resistance to corrosion, and have high surface area to volume ratio.[8] The solubility, processability, and stability are the foremost disadvantage of CPs, which somewhat limits their applications.[9] These drawbacks of CPs can be overcome by the formation of copolymer, hybrid, composite, interpenetrating network, etc.[10] The CPs are insulators at room temperature and the conductivity of these polymers can be increased by means of doping. Doping is a process by which donor/acceptor substituent are introduced into the polymer matrix, which leads to the formation of charged defects (polaron, bipolaron, and soliton) that work as a charge carrier and an increase in conductivity is observed. Since, most of the organic polymers do not have charge carriers, the requisite charge carriers can be introduced by the partial oxidation (p-doping) or reduction (n-doping) of the polymeric chain.[11]

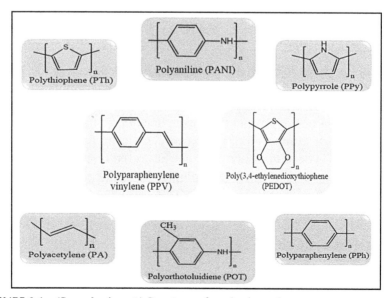

FIGURE 6.1   (**See color insert.**) Structures of conducting polymers.

Carbon nanotubes (CNTs) are well-ordered, hollow, sp^2 hybridized carbon materials having robust mechanical strength and unique morphology.[12] Since, their discovery (1991) by Sumio Iijima,[13] they have attracted wide attention to the researcher due to its high surface area, good electrical conductivity, and superior chemical stability, which makes them a superior candidate for many applications such as supercapacitors, sensors, batteries, etc.[14–16] Further, the high electrical conductivity of CNTs is due to its high mobility, low band gap, and long mean free path.[17] Both the CP and CNT exhibit a strong interfacial interaction due to its π–π interaction. The synthesized CNT has carbonaceous impurity and this impurity can be removed by the introduction of polymer.[18] Therefore, the CNT-based CPs offers an enhancement in electrical conductivity, surface area, mechanical strength and flexibility, which further enhanced its diverse applications shown in Figure 6.2. In addition to this, it also overcomes the shortcomings associated with the pristine CNT and CP.[19–21] Many attempts have been made to synthesize a mechanically robust, highly conducting, and flexible CNT-based CP composite material.

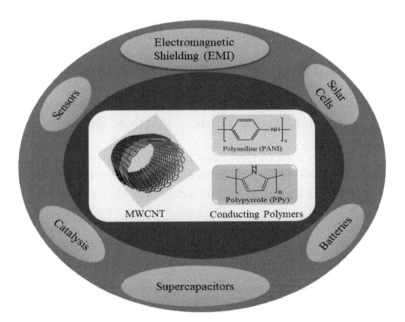

FIGURE 6.2 (See color insert.) Applications of CP/CNT composites.

Therefore, the present book chapter highlights the recent development in the CNT-based CP composites and their properties along with their diverse applications. In addition to this, future prospects of such materials have also been discussed.

6.2   CNT-BASED CP COMPOSITES

6.2.1   PANI/CNT COMPOSITE

Polyaniline (PANI)/CNT composites are the most widely studied composite so far as they offer enhanced electrical, optical, and chemical properties. For instance, free-standing three-dimensional (3-D) polyaniline–CNT/Ni-fiber hybrid electrodes have been synthesized by Li and coworkers who studied their electrochemical supercapacitive performance. The authors found that the hybrid electrode showed an outstanding specific capacitance of 725 F/g at a current density of 0.5 A/g and high energy density of 22 Wh/kg along with good cyclability. The interaction between the PANI and CNT provides large surface area, which is one of important characteristic for supercapacitor application. The Ni fiber acted as a charge collector while CNT helps to link that charge and the PANI acted as a charge promoter in the electrode.[22] Bang et al.[23] have synthesized a PANI/CNT composite film and studied its sensor behavior. It was observed that the composite film exhibits superior sensitivity and found to be efficient for near-infrared bolometric sensor. The significant enhancement in sensitivity was attributed to the synergistic effect of polyaniline and CNT, which further benefits the higher heat generation of CNTs and superior temperature coefficient of resistance of PANI. Later on, Cao et al.[19] have done a comparative study between Fe_3O_4/multiwalled carbon nanotubes (MWCNTs) and PANI/Fe_3O_4/MWCNTs composites. The authors found that the PANI/Fe_3O_4/MWCNTs composites showed better microwave absorption properties than that of Fe_3O_4/MWCNTs composite. A uricase-modified MWCNT (uricase-MWCNT)/PANI nanocomposite has been synthesized by Arora et al.[24] The authors found that the nanocomposite electrodes was highly sensitive and can detect a very low amount (0.01–1.0 mM) of uric acid. This is because of the loading of enzyme synergistically improved the electrochemical performance of the electrode. Ansari et al.[25] have fabricated MWCNT/graphene oxide (GO)/PANI composite and studied its photocatalytic against Cr (IV) and Congo red (CR) dye. It was observed that the maximum adsorption of Cr (IV) and CR was observed in an acidic medium at 30°C. Further, the kinetics for the adsorption behavior was also studied and it was found that the adsorption capacity was mainly dependent on the adsorbate concentration, solution pH, reaction temperature, and contact time. Further, Shen and coworkers[26] have fabricated a PANI/MWCNT nanocomposite and studied its electrical and mechanical properties for their application in electroconductive papers. They found that the incorporation of PANI/MWCNT nanocomposite to the cellulosic papers enhanced the electroconductivity (from 10^{-10} to 0.14 S/cm) and mechanical strength (29.99–38.15 N m/g) of the paper. Jo et al.[27] have synthesized a ternary hybrid composite using graphene, CNT, and PANI (GN/CNT/PANI) and studied its electrochemical properties for supercapacitor. The authors found that the synthesized composited showed an enhanced specific capacitance of 456 F/g and rate capability of 89%. The authors also found that the composite exhibits good stability and cyclability (97%) after 1000 cycles. The CNT acted as charge linkers between GN sheets, and PANI helps in the charge immobilization within the surface. A detailed study of CNT-based PANI composites is demonstrated in Table 6.1.

6.2.2   PPy/CNT COMPOSITE

Lu and coworkers[34] have synthesized a PPy/GN/CNT composite and studied their electrochemical properties. They found that the composite exhibits superior capacitive properties having specific capacitance of 361 F/g at a current density of 0.2 A/g, which was much higher than that of pristine PPy (176 F/g) and PPy/CNT (253 F/g); in addition to this, the composite showed only 4% capacitive loss after 2000 cycles. The enhanced capacitive properties of the composite were attributed to the synergistic properties of GN and CNT. Wang et al.[35] have synthesized a ternary composite (PPy/S-CNT) using PPy, sulfur (S), and CNT for their application in lithium batteries. The synthesized composite showed robust capacity of 600 mAh/g after 40 cycles, which was much higher

TABLE 6.1 Synthesis Method, Morphology, and Particle Size of the PANI/CNT Composites.

Composite system	Synthesis method	Morphology	Particle size (nm)	References
PANI/CNT	Interfacial polymerization	Nanotubes	10–70	[14]
PANI/CNT	Interfacial polymerization	Capped	–	[28]
PANI/uricase MWCNT	Electropolymerization	Globules	–	[24]
MWCNT/GO/PANI	Oxidative polymerization	Fibrous intermingled	–	[25]
G/CNT/PANI	Aerosol spray process	Crumpled	824	[27]
PANI/CNT/NiFe$_2$O$_4$	Emulsion polymerization	Needlelike	41	[29]
PANI/CNT	Solution casting	Uniform and smooth	27	[30]
PANI/CNT	Electropolymerization	Fiber-like	100	[31]
MnO$_2$/PANI/CNT	In situ polymerization	–	100	[32]
PANI/CNT	Surface-initiated polymerization	Core–shell	5–120	[33]

than that of binary (S-CNT) composite having capacity of 430 mAh/g. The enhancement in capacity was due to the PPy coating, which acted as conductive additive as well as an active material for charge transport. Later on, Sahmetlioglu et al.[36] have done extraction of Pb (II) from water sample using PPy and MWCNT as an active material. The authors found that the composite showed a satisfactory result with a minimum detection limit of 1.1 mg/L and preconcentration factors of 200, respectively. The adsorption capacity of the nanocomposite was found to be 25.0 mg/g, which reveals that the composite can detect Pb (II) from water sample. Roh and coworkers have synthesized an electrode material for microbial fuel cell using PPy/CNT composite. The composite electrode showed an improved performance with 38% more power density (287 mW/m^2) in comparison to that of pure CF electrode (208 mW/m^2). It was found that the CNT-based modification further improved their electron transfer ability by reducing the interfacial resistance.[37] Mahdavi et al. have fabricated a PPy/CNT nanocomposite and studied its antifouling property. The nanocomposite showed good antifouling properties as the surface roughness of membrane affects the fouling resistance. This was better proved by permeation test as the composite materials have high flux (96.9 L/m^2 h) as compared to the bare membrane (45.2 L/m^2 h), which confirms its efficiency as antifouling film.[38] Recently, Ramesh et al.[39] have synthesized a ternary hybrid nanocomposite (Co_3O_4/PPy/CNT) for their application in supercapacitor. The nanocomposite exhibits excellent capacitive behavior with a specific capacitance of 609 F/g at a current density of 3 A/g. the nanocomposite also displays high energy density of 84.58 Wh/kg and power density of 1500 W/kg along with its outstanding capacitance retention (97.1%) after 5000 cycles. This confirms its potential in the field of supercapacitor. A detailed study of CNT-based PPy composites is shown in Table 6.2.

6.2.3 PTh/CNT AND POLY(3,4-ethylenedioxythiophene)/ CNT COMPOSITE

Wang et al.[47] have fabricated a PTh/MWCNT composite by three different route (in situ polymerization, solution mixing, and ball milling) and studied its thermoelectric properties. The composite synthesized by solution mixing showed best thermoelectric properties. Further, the effect MWCNT was studied and found that the highest figure of merit (ZT) values (8.71×10^{-4}) were achieved when the MWCNT content was

TABLE 6.2   Synthesis Method, Morphology, and Particle Size of the PPy/CNT Composites.

Composite system	Synthesis method	Morphology	Particle size (nm)	References
PPy/CNT	Solution mixing	Sandwich	–	[40]
PPy/MWCNT	Dip and dry	Cauliflower	–	[41]
PPy/CNT	Electropolymerization	Fiber	30–70	[42]
PPy/CNT/MnO$_2$	Electropolymerization	Core-shell	25–55	[43]
PPy/SWCNT	Reflux technique	Fibrous	–	[44]
GN/PPy/CNT	In situ polymerization	Lamellar	2–100	[34]
PPy/GO/MWCNT	In situ polymerization	Uniform and Smooth	–	[45]
PPy/CNT	Electrodeposition	Porous	–	[46]

80%. Hong et al.[48] have synthesized a ternary $BaFe_{11.92}(LaNd)_{0.04}O_{19}$-titanium dioxide/MWCNT/PTh (BF-TD/MWCNTs/PTh) composite and studied its microwave absorption properties. The synthesized composite displayed good microwave absorption property when the amount of MWCNTs was 20.0 wt% and the ratio of the BF to TD was 4:5. A core–shell PTh/MWCNT nanocomposite has been synthesized by Swathy and coworkers using sodium bis(2-ethylhexyl) sulfosuccinate. The synthesized composites showed an increase in conductivity (1.5 times), which was much higher than that of pristine PTh. The higher conductivity was achieved due to their homogeneous mixing, which leads to the formation of well-arranged structure. Hence, an increase in conductivity was observed.[49] Chen et al.[50] have synthesized a 3-D CNT/poly(3,4-ethylenedioxythiophene) (PEDOT):polystyrene sulfonate (PSS) hydrogel composite for high-performance battery electrode. The synthesized composite hydrogels showed enhanced electrochemical properties. The 3-D composite hydrogel provides high conductivity and good mechanical strength and helps in the better ions transport, which are necessary for battery electrode. Later on, Jiang et al.[51] synthesized ternary (MWCNT/$Ni(OH)_2$/PEDOT:PSS) hybrid electrode for supercapacitor. The hybrid composite showed superior specific capacitance of 3262 F/g at 5 mV/s and retains 71.9% of its total capacitance at 100 mV/s. The MWCNT acted as a template for the deposition of amorphous $Ni(OH)_2$, which provides high surface area and excellent conductivity. Hu et al.[52] have fabricated a CNT/PEDOT composite and studied its thermoelectric properties. The composite film synthesized using two different solvent (hexane and xylene) showed an outstanding mechanical flexibility upon bending and twisting. The composite film exhibits a cable-like morphology with an excellent thermoelectric property than that of pure PEDOT synthesized in hexane. The synthesized composited showed a maximum power factor of 19.00 ± 1.43

$W/m/K^2$, which was 124 times greater than that of pristine PEDOT. A detailed study of CNT-based PTh and PEDOT composites is presented in Table 6.3.

TABLE 6.3 Synthesis Method, Morphology, and Particle Size of the PTh/CNT and PEDOT/CNT Composites.

Composite system	Synthesis method	Morphology	Particle size (nm)	References
PTh/MWCNT	Electropolymerization	Discrete spheroidal	2–3	[53]
PEDOT:PSS/CNT	Chemical vapor deposition	Flake-like	–	[54]
P3HT/SWNT	–	Mesh-like	–	[55]
PEDOT:PSS/ MWCNT	Solution casting	Uniform and smooth	–	[56]
PEDOT/CNT	Electropolymerization	Tubular	19–38	[57]
PEDOT:PSS/PVDF/ PEDOT:PSS/CNT	Spray coating	Rough and porous	–	[58]

6.3 PROPERTIES OF CP/CNT COMPOSITES

The combination of CPs and CNT provides unusual properties of both the component and exhibit high surface area, excellent electrical, thermal, and mechanical properties. These properties further depend on certain factors such as size, shape, aspect ratio, dispersion, and alignment of the filler within the matrix.[59] The electrical, thermal, and mechanical properties of the composite increases with the addition of filler particle but these are mainly dependent on the homogeneous dispersion of filler. Homogeneous dispersion of filler leads to the better interaction between filler and matrix, which further enhances its thermal and mechanical stability,[60] whereas in case of electrical property, homogeneous dispersion leads to the better alignment in the molecular structure, which further helps in the transportation of electrons within the molecule, hence an increase in conductivity is observed.[26] A detail study of thermal, electrical, and mechanical properties of CP/CNT composites is reported in Table 6.4.

TABLE 6.4 Thermal, Electrical, and Mechanical Properties of CP/CNT Composites.

CP/CNT composite	Thermal property	Electrical property	Mechanical property	References
PANI/CNT	—	Bulk conductivity was found to be 10–6000 S/cm, while the sheet resistance was 300 Ω/cm	Showed good mechanical stability up to 100 bend cycle	[14]
PTh/MWCNT	Thermal properties of the composites increase with the increase in amount of MWCNT. The initial thermal degradation was observed at 230 °C with below 50 wt% MWCNT content, while it increases to 270 °C with 60 wt% MWCNT content. These results are quite higher than that of pristine PTh (190°C)	The room temperature electrical conductivity was found to be in the range of 2.34–2982 S/m with the increase in amount of MWCNT from 10 to 90 wt%. the highest conductivity of 2982 S/m was observed with maximum loading (90 wt%) of MWCNT, which was 11 times higher than that of pure PTh	—	[60]
BF-TD/MWCNTs/PTh	Pure Pth showed initial thermal degradation below 110°C, whereas in case of composite, it was 230°C, which was much higher than that of pure PTh. This clearly reveals that the composite has higher thermal stability as compared to neat PTh	The conductivity was found to increase with the increase in amount of MWCNT. The highest conductivity of 0.92 S/cm was achieved with the loading of 20 wt% MWCNT	—	[48]
TiO2–PEDOT:PSS–CNT	—	The composite film exhibits a conductivity of 215 S/cm, which is found to be lower than that of PEDOT:PSS–CNT film having conductivity of 950 S/cm	The composite film showed insignificant change after 500 bending cycles along with a bending radius of 3.5 mm. This confirms its mechanical stability	[50]

TABLE 6.4 *(Continued)*

CP/CNT composite	Thermal property	Electrical property	Mechanical property	References
PANI/CNT	In case of pure PANI, total weight loss of 69.31% occurs at 600°C, whereas with the increase in concentration of CNT (1%), only 55.34% weight loss occur in the same temperature. The same trend was noticed with different amount of (3% and 5%) of CNT and a total weight loss of 52.16% and 46.93% was observed	The electrical conductivity of PANI/CNT composite was found to increase with the increase in CNT content (1%, 3%, and 5%). The maximum conductivity of 2.4 S/cm was observed in case of PANI/CNT (5%) composite	PANI showed a good mechanical strength of 1.91 Gpa, whereas the PANI/CNT (1%) composites showed superior mechanical strength of 2.3 Gpa, which was much higher than that of pristine PANI. Further, the mechanical strength was found to increase with the increase in CNT content	[61]
PANI/MWCNT	—	The PANI/MWCNT (20%) composite showed an electrical conductivity of 0.14 S/m, which was much higher than that of pure PANI 10^{-10} S/m	The mechanical strength of the composite was found to increase as the MWCNT content increases and have a tensile strength of 38.15 Nm/g, which was quite higher than that of pristine PANI (29.99 Nm/g)	[26]

6.4 FUTURE PROSPECTS AND CONCLUSIONS

Owing to their superior electrical, thermals and mechanical properties, the CP/CNT composites have gained much attention. Further, they have been widely used in many applications such as supercapacitors, sensors, batteries, EMI shielding, etc. The CP/CNT composites have been widely used in the field of supercapacitor and many better results are observed, but there is still lot of scope of research as the researchers have failed to make highly flexible and thermally stable supercapacitor. Further, these materials have shown their high efficiency in the field of sensor. The future development should focus on the development of highly sensitive and more efficient sensor having excellent selectivity and response time. Therefore, the future development should focus on this field. In case of battery, the composite material was found to be quite efficient but the major concern of these materials is their stability. Therefore, future development should focus on highly stable and efficient battery cathode material.

ACKNOWLEDGMENTS

The author, Sajid Iqbal is highly thankful to UGC Research Fellowship and Department of Chemistry, Jamia Millia Islamia, New Delhi, India. Dr. Anujit Ghosal is thankful to Government of India, Science and Engineering Research Board (SERB) for financial support in the form of National post-doctoral Fellowship (PDF/2016/003866). The authors are also thankful to the School of Biotechnology, Jawaharlal Nehru University, New Delhi 110067, India for implementation of fellowship.

KEYWORDS

- nanocomposite hybrids
- doping
- synthetic metals
- multiwall carbon nanotubes
- polyaniline

REFERENCES

1. Weinberger, B. R.; Kaufer, J.; Heeger, A. J.; Pron, A.; MacDiarmid, A. G. *Phys. Rev. B* **1979**, *20* (1), 223–230. DOI: 10.1103/PhysRevB.20.223.
2. Chen, B.; Liu, C.; Ge, L.; Hayashi, K. *Sensors Actuators, B Chem.* **2017**, *241*, 1099–1105. DOI: 10.1016/j.snb.2016.10.030.
3. Ke, F.; Liu, Y.; Xu, H.; Ma, Y.; Guang, S.; Zhang, F.; Lin, N.; Ye, M.; Lin, Y.; Liu, X. *Compos. Sci. Technol.* **2017**, *142*, 286–93. DOI: 10.1016/j.compscitech.2017.02.026.
4. Fan, Q.; Su, W.; Guo, X.; Guo, B.; Li, W.; Zhang, Y.; Wang, K.; Zhang, M.; Li, Y. *Adv. Energy Mater.* **2016**, *6* (14). DOI: 10.1002/aenm.201600430.
5. Yan, J.; Huang, Y.; Chen, X.; Wei, C. *Synth. Met.* **2016**, *221*, 291–298. DOI: 10.1016/j.synthmet.2016.09.018.
6. Rawat, N. K.; Ghosal, A.; Ahmad, S. *RSC Adv.* **2014**, *4* (92), 50594–50605. DOI: 10.1039/C4RA06679K.
7. Zhou, Q.; Shi, G. *J. Am. Chem. Soc.* **2016**, *138* (9), 2868–2876. DOI: 10.1021/jacs.5b12474.
8. Khatoon, H; Ahmad, S. *J. Ind. Eng. Chem.* **2017**, *53*, 1–22.
9. Jain, S.; Karmakar, N.; Shah, A.; Kothari, D. C.; Mishra, S.; Shimpi, N. G. *Appl. Surf. Sci.* **2017**, *396*, 1317–1325. DOI: 10.1016/j.apsusc.2016.11.154.
10. Ghosal, A.; Rahman, O. U.; Ahmad, S., *Indus. Eng. Chem. Res.* **2015**, *54* (51), 12770–12787.
11. Ziadan, K. M. *New Polymers for Special Applications*, InTech Publishing: London, UK, 2012; pp 3–24.
12. Yang, Q.; Pang, S. K.; Yung, K. C. *J. Electroanal. Chem.* **2014**, *728*, 140–147. DOI: 10.1016/j.jelechem.2014.06.033.
13. Iijima, S. *Nature* **1991**, *354*, 56–58.
14. Salvatierra, R. V.; Cava, C. E.; Roman, L. S.; Zarbin, A. J. G. *Adv. Funct. Mater.* **2013**, *23* (12), 1490–1499. DOI: 10.1002/adfm.201201878.
15. Moniruzzaman, M.; Sahoo, S.; Ghosh, D.; Das, C. K.; Singh, R. *J. Appl. Polym. Sci.* **2013**, *128* (1), 698–705. DOI: 10.1002/app.38202.
16. Hu, X.; Chen, L.; Tan, L.; Ji, T.; Zhang, Y.; Zhang, L.; Zhang, D.; Chen, Y. *J. Mater. Chem. A* **2016**, *4* (17), 6645–6652. DOI: 10.1039/C6TA00287K.
17. Kim, S. H.; Kim, S. Y.; Shin, U. S. *Compos. Sci. Technol.* **2016**, *126*, 78–85. DOI: 10.1016/j.compscitech.2016.02.017.
18. Mylvaganam, K.; Zhang, L. C. *Recent Pat. Nanotechnol.* **2007**, *1*, 59–65.
19. Arti, V.; Ajeet, K.; Atul, V.; Vidya, S.; Anujit, G.; K., G. Y.; Sharif, A.; Madhavan, N. *Adv. Healthcare Mater.,* **2018**, *7* (9), 1701213.
20. Gupta, T. K.; Singh, B. P.; Mathur, R. B.; Dhakate, S. R. *Nanoscale* **2014**, *6* (2), 842–851. DOI: 10.1039/C3NR04565J.
21. Bin Yang, R.; Reddy, P. M.; Chang, C. J.; Chen, P. A.; Chen, J. K.; Chang, C. C. *Chem. Eng. J.* **2016**, *285*, 497–507. DOI: 10.1016/j.cej.2015.10.031.
22. Li, Y.; Fang, Y.; Liu, H.; Wu, X; Lu, Y. *Nanoscale* **2012**, *4*, 2867.
23. Bang, D.; Lee, J.; Park, J.; Choi, J.; Chang, Y. W.; Yoo, K; Bang, D.; Lee, J.; Park, J.; Choi, J.; Chang, Y. W.; Yoo, K. H.; Huh, Y. M.; Haam, S. H. *J. Mater. Chem.* **2012**, *22*, 3215.

24. Arora, K.; Choudhary, M.; Malhotra, B. D. *Appl. Biochem. Biotechnol.* **2014,** *174* (3), 1174–1187. DOI: 10.1007/s12010-014-0996-x.

25. Ansari, M. O.; Kumar, R.; Ansari, S. A.; Ansari, S. P.; Barakat, M. A.; Alshahrie, A.; Cho, M. H. *J. Colloid Interface Sci.* **2017,** *496* (Vi), 407–15. DOI: 10.1016/j.jcis.2017.02.034.

26. Shen, X.; Tang, Y.; Zhou, D.; Zhang, J.; Guo, D.; Friederichs, G. *J. Bioresour. Bioprod.* **2016,** *1* (1), 48–54.

27. Jo, E. H.; Jang, H. D.; Chang, H.; Kim, S. K.; Choi, J. H.; Lee, C. M. *ChemSusChem* **2017,** *10* (10), 2210–2217. DOI: 10.1002/cssc.201700212.

28. Mariano, L. C.; Salvatierra, R. V.; Cava, C. E.; Koehler, M.; Zarbin, A. J. G.; Roman, L. S. *J. Phys. Chem. C* **2014,** *118* (43), 24811–24818. DOI: 10.1021/jp502650u.

29. Sabet, M.; Jahangiri, H.; Ghashghaei, E. *Synth. Met.* **2017,** *224*, 18–26. DOI: 10.1016/j.synthmet.2016.11.034.

30. Saadattalab, V.; Shakeri, A.; Gholami, H. *Prog. Nat. Sci. Mater. Int.* **2017,** *26* (6), 517–522. DOI: 10.1016/j.pnsc.2016.09.005.

31. Pan, S.; Yang, J. *J. Nanosci. Nanotechnol.* **2015,** *15*, 7412–7415.

32. Wenjie Liu, X.; Wang, S.; Wu, Q.; Huan, L; Zhang, X.; Chao, Y.; Chen, M. *Chem. Eng. Sci.* **2016,** *156*, 178–185.

33. Ji, T.; Tu, R.; Mu, L.; Lu, X.; Zhu, J. *Appl. Catal. B Environ.* **2018,** *220*, 581–588. DOI: 10.1016/j.apcatb.2017.08.066.

34. Lu, X.; Zhang, F.; Dou, H.; Yuan, C.; Yang, S.; Hao, L.; Shen, L.; Zhang, L.; Zhang, X. *Electrochim. Acta* **2012,** *69*, 160–166. DOI: 10.1016/j.electacta.2012.02.107.

35. Wang, J.; Lu, L.; Shi, D.; Tandiono, R.; Wang, Z.; Konstantinov, K.; Liu, H. *Chempluschem* **2013,** *78* (4), 318–324. DOI: 10.1002/cplu.201200293.

36. Sahmetlioglu, E.; Yilmaz, E.; Aktas, E.; Soylak, M. *Talanta* **2014,** *119*, 447–451. DOI: 10.1016/j.talanta.2013.11.044.

37. Roh, S-H.; Woo, H-G. *J. Nanosci. Nanotechnol.* **2015,** *15* (1), 484–487. DOI: 10.1166/jnn.2015.8404.

38. Mahdavi, M. R.; Delnavaz, M.; Vatanpour, V.; Farahbakhsh, J. *Sep. Purif. Technol.* **2017,** *184*, 119–127. DOI: 10.1016/j.seppur.2017.04.037.

39. Ramesh, S.; Haldorai, Y.; Kim, H. S.; Kim, J-H. *RSC Adv.* **2017,** *7* (58), 36833–36843. DOI: 10.1039/C7RA06093A.

40. Liang, L.; Chen, G.; Guo, C. Y. *Compos. Sci. Technol.* **2016,** *129*, 130–136. DOI: 10.1016/j.compscitech.2016.04.023.

41. Liu, C.; Cai, Z.; Zhao, Y.; Zhao, H.; Ge, F. *Cellulose* **2016,** *23* (1), 637–648. DOI: 10.1007/s10570-015-0795-8.

42. Yesi, Y.; Shown, I.; Ganguly, A.; Ngo, T. T.; Chen, L. C.; Chen, K. H. *ChemSusChem* **2016,** *9* (4), 370–378. DOI: 10.1002/cssc.201501495.

43. Li, P.; Yang, Y.; Shi, E.; Shen, Q.; Shang, Y.; Wu, S.; Wei, J.; Wang, K.; Zhu, H.; Yuan, Q.; Cao, A.; Wu, D. *ACS Appl. Mater. Interfaces* **2014,** *6* (7), 5228–5234. DOI: 10.1021/am500579c.

44. He, B.; Tang, Q.; Luo, J.; Li, Q.; Chen, X.; Cai, H. *J. Power Sour.* **2014,** *256*, 170–177. DOI: 10.1016/j.jpowsour.2014.01.072.

45. Wang, B.; Qiu, J.; Feng, H.; Sakai, E. *Electrochim. Acta* **2015,** *151*, 230–239. DOI: 10.1016/j.electacta.2014.10.153.

46. Li, J.; Wojtal, P.; Wallar, C. J.; Zhitomirsky, I. *Mater. Manuf. Process.* **2017,** 6914 (September), 1–5. DOI: 10.1080/10426914.2017.1364850.

47. Wang, L.; Jia, X.; Wang, D.; Zhu, G.; Li, J. *Synth. Met.* **2013,** *181,* 79–85. DOI: 10.1016/j.synthmet.2013.08.011.

48. Hong, X.; Xie, Y.; Wang, X.; Li, M.; Le, Z.; Gao, Y.; Huang, Y.; Qin, Y.; Ling, Y. *Compos. Sci. Technol.* **2015,** *117,* 215–224. DOI: 10.1016/j.compscitech.2015.06.022.

49. Swathy, T. S.; Jose, M. A.; Antony, M. J. *Polym.* (United Kingdom) **2016,** *103,* 206–213. DOI: 10.1016/j.polymer.2016.09.047.

50. Chen, Z.; To, J. W. F.; Wang, C.; Lu, Z.; Liu, N.; Chortos, A.; Pan, L.; Wei, F.; Cui, Y.; Bao, Z. *Adv. Energy Mater.* **2014,** *4* (12). DOI: 1–10. 10.1002/aenm.201400207.

51. Jiang, W.; Yu, D.; Zhang, Q.; Goh, K.; Wei, L.; Yong, Y.; Jiang, R.; Wei, J.; Chen, Y. *Adv. Funct. Mater.* **2015,** *25* (7), 1063–1073. DOI: 10.1002/adfm.201403354.

52. Hu, X.; Chen, G.; Wang, X. *Compos. Sci. Technol.* **2017,** *144,* 43–50. DOI: 10.1016/j.compscitech.2017.03.018.

53. Zhang, H.; Hu, Z.; Li, M.; Hu, L.; Jiao, S. *J. Mater. Chem. A* **2014,** *2* (40), 17024–17030. DOI: 10.1039/c4ta03369h.

54. Zhou, Y.; Lachman, N.; Ghaffari, M.; Xu, H.; Bhattacharya, D.; Fattahi, P.; Abidian, M. R.; Wu, S.; Gleason, K. K.; Wardle, B. L.; Zhang, Q. M. *J. Mater. Chem. A* **2014,** *2* (26), 9964–9969. DOI: 10.1039/c4ta01785d.

55. Habisreutinger, S. N.; Leijtens, T.; Eperon, G. E.; Stranks, S. D.; Nicholas, R. J.; Snaith, H. J. *Nano Lett.* **2014,** *14* (10), 5561–5568. DOI: 10.1021/nl501982b.

56. Sharma, S.; Hussain, S.; Singh, S.; Islam, S. S. *Sensors Actuators, B Chem.* **2014,** *194,* 213–219. DOI: 10.1016/j.snb.2013.12.050.

57. Braik, M.; Barsan, M. M.; Dridi, C.; Ben Ali, M.; Brett, C. M. A. *Sensors Actuators, B Chem.* **2016,** *236,* 574–582. DOI: 10.1016/j.snb.2016.06.032.

58. Simaite, A.; Delagarde, A.; Tondu, B.; Souères, P.; Flahaut, E.; Bergaud, C. *Nanotechnology* **2017,** *28* (2), 25502. DOI: 10.1088/0957-4484/28/2/025502.

59. Meer, S.; Kausar, A.; Iqbal, T. *Polym. Plast. Technol. Eng.* **2016,** *55* (13), 1416–1440. DOI: 10.1080/03602559.2016.1163601.

60. Qian, T.; Zhou, X.; Yu, C.; Wu, S.; Shen, J. *J. Mater. Chem. A* **2013,** *1,* 15230–15234. DOI: 10.1039/C3TA13624H.

61. Kumar, A.; Kumar, V.; Awasthi, K. *Polym. Plast. Technol. Eng.* **2017,** *26,* 1–28. DOI: 10.1080/03602559.2017.1300817.

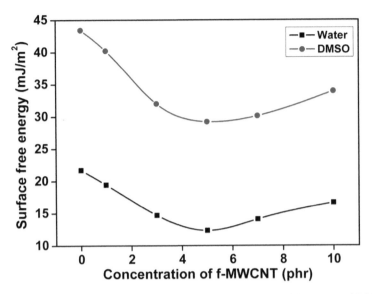

FIGURE 3.3   Surface energy plots of SBR nanocomposites with water and DMSO as a function of filler concentration.

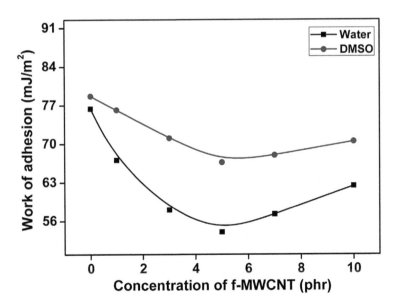

FIGURE 3.4   Work of adhesion of SBR nanocomposites with water and DMSO as a function of filler concentration.

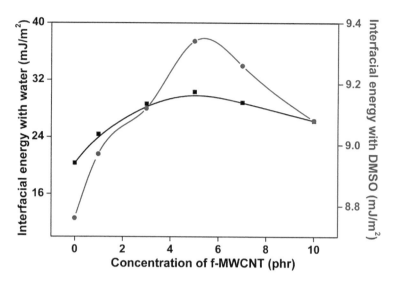

FIGURE 3.5 Interfacial energy of SBR nanocomposites with water and DMSO as a function of filler concentration.

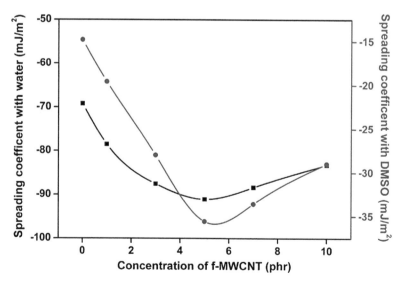

FIGURE 3.6 Spreading coefficients of SBR nanocomposites with water and DMSO as a function of filler concentration.

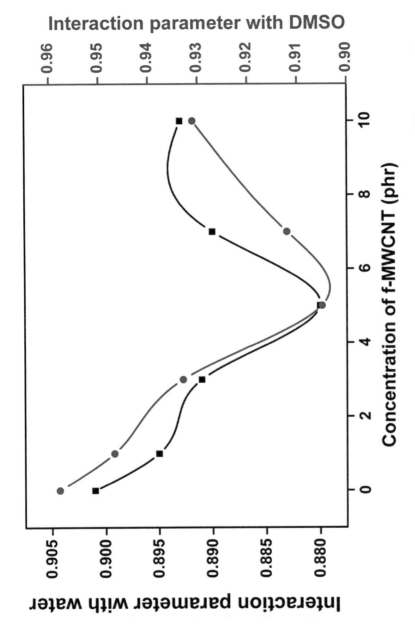

FIGURE 3.7 Interaction parameters of SBR nanocomposites with water and DMSO as a function of filler concentration.

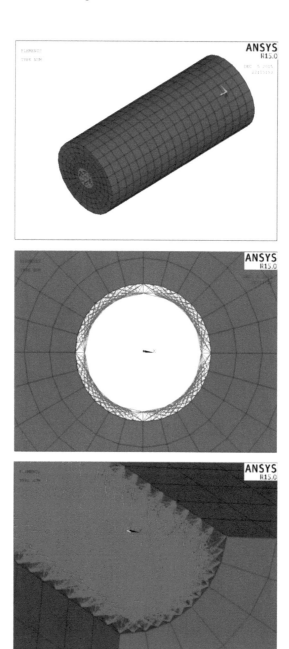

FIGURE 4.5   Three parts of RVE from different point of view.

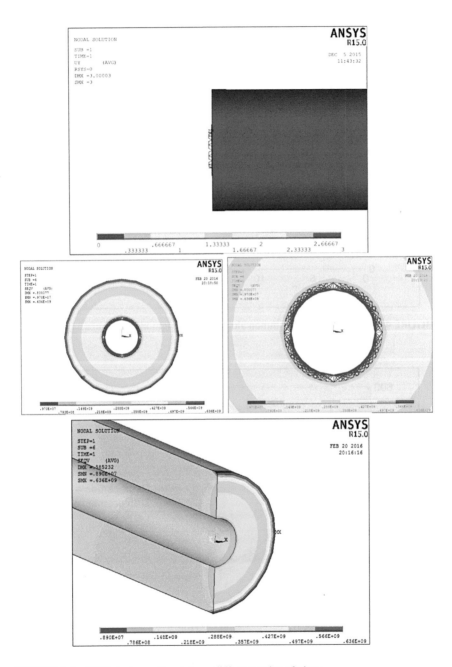

FIGURE 4.6   RVE during pull-out from different point of view.

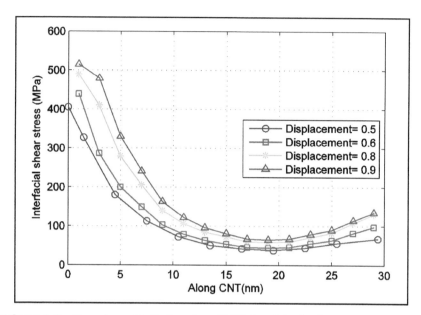

FIGURE 4.8   Shear stress distribution along CNT's length in elastic area.

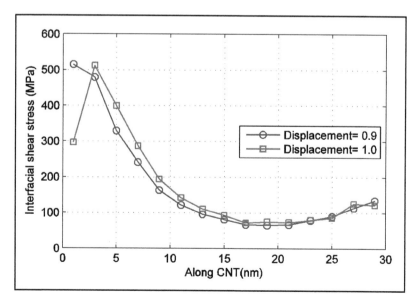

FIGURE 4.9   Shear stress distribution along CNT's length at the end of elastic area (displacement 0.9 nm) and initial damage area (displacement 1 nm).

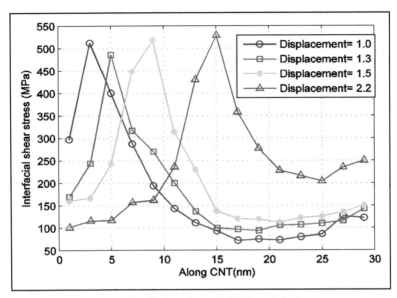

FIGURE 4.10   Shear stress distribution along CNT's length in damage area.

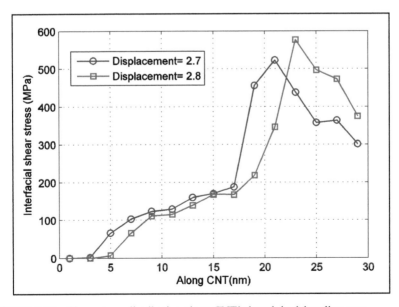

FIGURE 4.11   Shear stress distribution along CNT's length in debonding area.

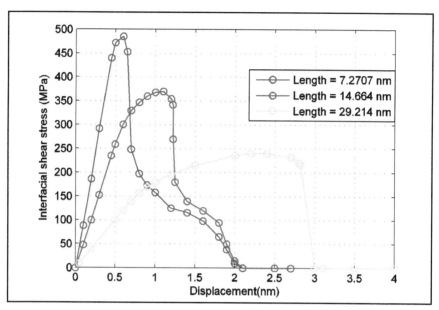

FIGURE 4.12   Shear stress diagram for different length shorter than 29.412 nm.

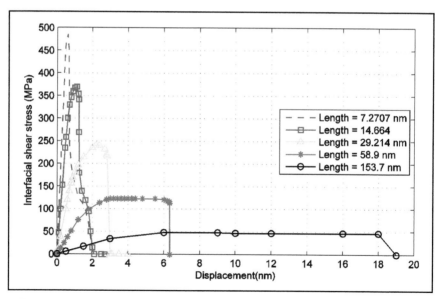

FIGURE 4.15   Shear stress diagram for different length from 7.2707 to 153.7 nm.

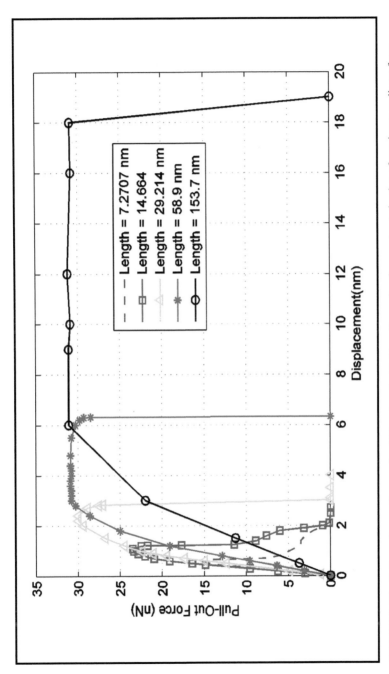

FIGURE 4.16 Pull-out forces diagram for different embedded length ranging from; saturated value of maximum pull-out force.

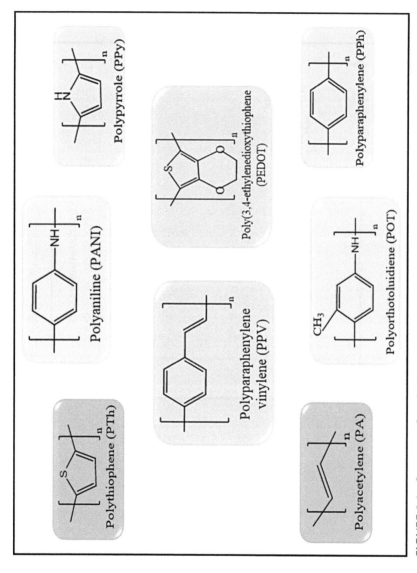

FIGURE 6.1 Structures of conducting polymers.

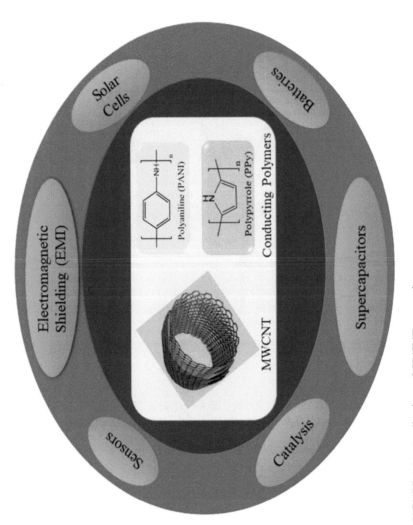

FIGURE 6.2 Applications of CP/CNT composites.

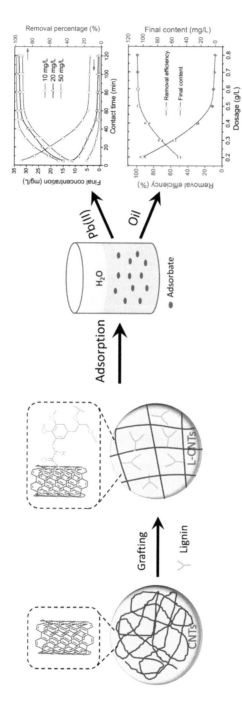

FIGURE 7.6 Adsorption of Pb (II) on lignin-grafted carbon nanotubes (L-CNTs) (Adapted from Li et al.,[85] Copyright (2017), with permission from Elsevier).

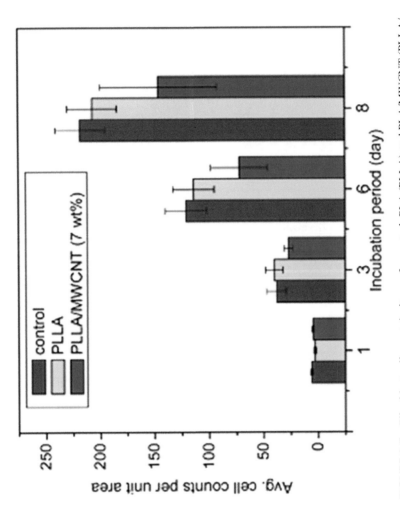

FIGURE 8.7 Fibroblast cell counts/unit area for control, PLA (PLLA) and PLA/MWCNT (PLLA/MWCNT) (7 wt%). Source: Reproduced from Ref. [29].

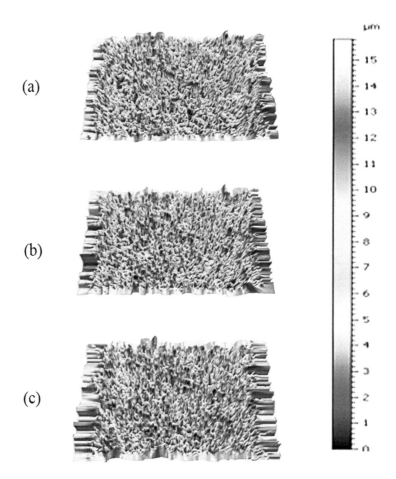

FIGURE 9.13 Surface topography in three different views: (a) 15°, (b) 30°, and (c) 45°.

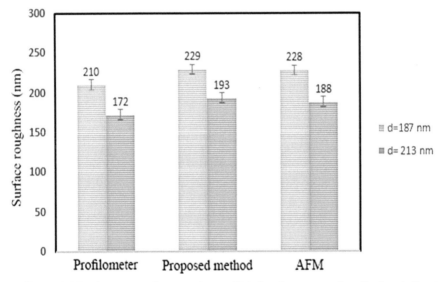

FIGURE 9.27 Average surface roughness (Ra) for the proposed method and direct measurement by profilometer and AFM.

CHAPTER 7

CARBON NANOTUBES-BASED ADSORBENT: AN EFFICIENT WATER PURIFICATION TECHNOLOGY

RANGNATH RAVI[1], SAJID IQBAL[1], ANUJIT GHOSAL[2*], and SHARIF AHMAD[1*]

[1]Materials Research Laboratory, Department of Chemistry, Jamia Millia Islamia, New Delhi 110025, India

[2]Plant Nanobiotechnology Laboratory, School of Biotechnology, Jawaharlal Nehru University, New Delhi 110067, India

*Corresponding authors.
E-mail: sharifahmad_jmi@yahoo.co.in; anuj.ghosal@gmail.com

ABSTRACT

Nanostructures of carbon-based materials and their composites have been found be the very important class of materials, due to their versatile properties in the field of biomedicine, engineering, optics, water purification, etc. The large surface area, intrinsic conductivity, ability to surface functionalization, biocompatibility, etc. are responsible for the cut-edge performance of carbon-based nanostructures. In past few decades, the carbon nanotubes (CNT), carbon nanodots (CND), graphene, graphite, fullerene (C_{60}), etc. have been efficiently synthesized in the laboratory as well as on industrial scale and have employed them for various applications. This chapter discusses the current status of different contaminant polluting the water system, their effects on human health, and their efficient removal. Out of various modes of water purification, here, advancements in CNTs-based composites have been elaborated. The chemical or biological treatments which can further improve their adsorption properties have also been presented.

7.1 INTRODUCTION

Reliable and sustainable water supply is the most essential and concerning subject for today's world; it remains a challenge to meet global water demands as 97.5% of natural water is saline and only 2.5% is fresh. Thus, the supply of fresh water to civilized people is very hard and even impossible task, as the population of the world is steadily increasing at an alarming rate. According to the rate of growth of population, more than 64 billion cubic meters of fresh water per year would be required. Other than this, due to rapid industrialization, agricultural activities, and geological and environmental changes, quality of water at water resources is deteriorating every day. Therefore, water pollution has become a serious issue in the present and for the future generations, affecting all living creatures, directly as well as indirectly. WHO, environmental agencies, government authorities, scientists, and academicians all over the world are working together to find a suitable solution for the issue of water contamination, its purification, and regeneration. Thousands of organic, inorganic, and biological species have been reported as water contaminants; they in general either produce the foul smell, impart color, or are found to be toxic to the living organism at particular concentration.[1-5] For example, pollutants such as heavy metals (Hg, As, Pb, etc.) contamination, and colored dyes used in leather industries, clothes, etc. have been causing serious adverse effects with lethal and carcinogenic activities once entered in the human biological system, thus, continuously damaging the ecosystem.[6,7] Various physical, chemical, and biological methodologies have been employed and have been continuously developed for remediation of this problem. Boiling of water has been most commonly employed way which can get you clean water from the biological point of view but not from inorganic solids and sometimes thermally stable organic molecules. Various chemical treatments can get us clean water but all of these are expensive and accompany a possibility of further polluting the ecosystem by the waste output of the process, whereas, adsorption processes has been found to have great potential from waste removal, regeneration of adsorbent, and utilization of pollutant obtained. Here, out of different pollutant CNT-based wastewater adsorbents have been profoundly discussed.

7.2 WATER POLLUTANTS

Pollutants can alter the natural quality of the water respond to change in physical, chemical, or biological properties. These pollutants can be characterized as physical, chemical, or biological pollutants. Physical pollutants may be larger particles such as discarded materials from human activities (e.g., plastic bags and bottles) or physical factors such as temperature and pH change. Although these materials are not so harmful to human health, they comprise the visual impact of water pollution. Chemical pollutants can be small atoms, molecules, or ions such as organic molecules such as dyes like methylene blue, Congo red (CR), and Rhodamine B, etc. inorganic ions such as NO_3^-, PO_4^{2-}, F^-, and SO_4^{2-} and heavy metals such as As^{3+}, As^{5+}, Pb^{2+}, Hg^{2+}, etc. which are anthropologically discharged into water sources and contaminate the water.[8-13]

For example, mercury emanating from mining activity, certain nitrogen compounds used in agriculture, chlorinated organic molecules arising from sewage or water treatment plants, or various acids which are the results of various manufacturing activities.[14,15] Radioactive substances are also characterized as chemical pollutants.[16] Most discharge of radioactivity is not from the negligible escape from nuclear power plants, but rather arises from agricultural practices such as tobacco farming, where radioactive contamination of phosphate fertilizer is a common method of introduction of radioactive materials into the environment. In addition, pathogenic microbes are living organisms such as virus, bacteria, fungi, protozoa, plankton, worms, and other small creatures characterized as

TABLE 7.1 Different Industries and Their Hazardous Wastes.

Products	Hazardous wastes
Medicines	Organic solvents and residues, heavy metals (mercury and zinc)
Paints	Heavy metals, pigments, organic dyes, solvents
Leather	Heavy metals organic solvents
Oil, petroleum products	Oils, phenols, organic compounds, heavy metals, etc.
Plastics	Organic chlorine compounds
Textiles	Heavy metals, dyes, organic chlorine compounds, and solvents
Pesticides	Organic chlorine compounds and organic phosphate compounds
Metals	Heavy metals, fluorides, cyanides, acids and alkaline cleaners, solvent and pigments

biological pollutants which are introduced into natural water bodies from untreated sewage or surface runoff from intensive livestock grazing and cause the several diseases (Table 7.1).[17]

7.3 EFFECTS OF WATER POLLUTION ON HUMAN HEALTH

Water pollutants are very harmful not only for the human body but also for animals and aquatic life as it has been reported that polluted water is a major cause of human disease, misery, and death.[18] According to the previous and recent research the nonbiodegradable pollutants such as heavy metals such as As (III), As (V), Pb (II), Cd (II) and Hg (II), etc. are most dangerous to the human life.[19,20] Nowadays, inorganic metals such as arsenic, cadmium, chromium, cobalt, copper, lead, manganese, mercury, nickel, and zinc metals are a major environmental concern. Certainly, that is the case of pollution of aquatic systems by heavy metals.[19–22]

Heavy metals are trace elements with higher specific gravity found in the earth crust, are nonbiodegradable, and bioaccumulates in the body. These metal ions have a tendency to bind to the food chain, sometimes, get stored in soft and hard tissues such as kidney, liver, and bones and thus a great threat to the health affecting growth and development of the organisms.[21,23] Besides, these metals show acute toxicity leading to cancers.[24] The acute toxicity of metal ions is because of their tendency to interact with several biological enzymes (Table 7.2).[25]

For example, arsenic is 20th most abundant element found in the earth crust and is associated with a large number of hazardous effects on human. On inhalation of arsenic contaminated water, it produces toxic intermediates during the metabolism which in high concentration do not rid from the body and bind to soft or hard tissue, where it gets absorbed. It replaces the phosphate group involved in various biological pathways, inhibits glucose transporters, alters expression of genes, and can stimulate oxidative stress.[40]

Generally, arsenic shows adverse effects through inducing the oxidative stress by affecting the naturally found antioxidants in the body. It stimulates the generation of reactive oxygen species (ROS) by the abnormal transfer of an electron through the respiratory organ to mitochondrion cell. These electrons trigger the molecular oxygen (O_2) to form superoxide anion (O_2^-) and then dismutate to H_2O_2 in the mitochondrial

TABLE 7.2   Effects of Heavy Metals on Human Health.

Heavy metals	Effects	References
Arsenic	Dermal, neurological, cardiovascular, mutagenic, renal, diabetes, hematological and carcinogenic effect	[26, 27]
Lead	Nausea, encephalopathy, headache and vomiting, learning difficulties, mental retardation, hyperactivity, vertigo, kidney damage, birth defects, muscle weakness, anorexia, cirrhosis of the liver, thyroid dysfunction, insomnia, fatigue, degeneration of motor neurons, schizophrenic-like behavior, various forms of blood disorders and anemia, rapid deterioration of brain and the nervous system, reduced fertility both in men and women, and Alzheimer disease	[28, 29]
Chromium	Skin ulcer, stomach upsets and ulcers, convulsions, skin rashes, respiratory problems, hemolysis, acute renal failure, weakened immune systems, kidney and liver damage, alteration of genetic material, lung cancer, and pulmonary fibrosis	[30, 31]
Cadmium	Renal disinfection, kidney damage, hypertension, neurological effect, bone effect, Itai-Itai disease, and cancer	[32, 33]
Copper	Diarrhea, renal failure, liver damage, food poisoning, hepatic injury, Wilson's disease, hypertension, nausea/vomiting, hyperactivity, postpartum psychosis, cardiovascular disease, mental disorders, anemia; arthritis/rheumatoid arthritis, inflammation and enlargement of liver, heart problem, pink disease, and cystic fibrosis	[34, 35]
Mercury	Tremors, birth defects, kidney or lungs damage, nausea, loss of hearing, gingivitis, chromosome damage, mental retardation, tooth loss, seizures, cerebral palsy, blindness and deafness, hypertonia—muscle rigidity, and minamata disease	[36, 37]
Nickel	Dermatitis, myocarditis, encephalopathy, pulmonary fibrosis, cancer of lungs, nose and bone, headache, dizziness, nausea/vomiting, chest pain, and rapid respiration	[38, 39]

cell. The production of H_2O_2 is the result of the generation of methylated arsenic metabolites radicals such as $(CH_3)_2$ As• and $CH_3)_2AsOO$•, during the oxidation of As^{3+} to As^{5+}. These free radicals are responsible for oxidative stress (Fig. 7.1).[41]

In addition to oxidative stress, arsenic shows the large binding affinity to thiol groups present in the enzymes and therefore, affect their catalytic activities. Arsenic forms a strong complex with vicinal thiol of enzymes

FIGURE 7.1 Arsenic methylation pathway in the human body. Where A–D are enzymes as—A: arsenate reductase or purine nucleoside phosphorylase (PNP), B: arsenite methyltransferase (As3MT), C: glutathione S-transferase omega 1 or 2 (GSTO1, GSTO2), D: arsenite methyltransferase (As3MT), MMAV: monomethylarsenic acid, MMAIII: monomethylarsonous acid, DMAV: dimethylarsenic acid.[41]

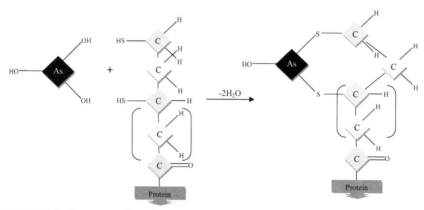

FIGURE 7.2 Representation of toxicity of arsenite (As^{3+}) via sulfhydryl–arsenic bonding.

and shows toxicity. It may decrease the cellular reduced glutathione (GSH) level in the body through bind to their sulfhydryl sites (Fig. 7.2).[42]

In addition to heavy metals, inorganic ions such as Na^+, K^+, Ca^{2+}, Mg^{2+}, SO_4^{2-}, Cl^-, CO_3^{2-}, PO_4^{3-}, NO_3^-, NO_2^-, Br^-, F^-, Li^+, and NH_4^+ are also found in water. The natural sources of inorganic ionic species in water systems include the weathering of rocks, atmospheric deposition, and groundwater runoffs, whereas anthropogenically, these ions inputs into water bodies from agriculture, animal husbandry, aquaculture activities, and municipal and industrial wastewaters.[19] List of essential elements and their necessary percentage in the body is given in Figure 7.3.

Some of the ions, such as chloride and carbonate, are necessary to live in proper concentrations, but others such as nitrates and nitrites can be dangerous to health at high concentrations.[43–45] Fluoride is essential to prevent dental caries; however, high level of fluoride causes dental

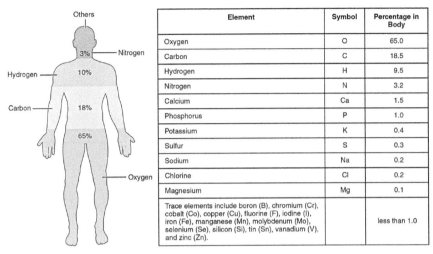

Element	Symbol	Percentage in Body
Oxygen	O	65.0
Carbon	C	18.5
Hydrogen	H	9.5
Nitrogen	N	3.2
Calcium	Ca	1.5
Phosphorus	P	1.0
Potassium	K	0.4
Sulfur	S	0.3
Sodium	Na	0.2
Chlorine	Cl	0.2
Magnesium	Mg	0.1
Trace elements include boron (B), chromium (Cr), cobalt (Co), copper (Cu), fluorine (F), iodine (I), iron (Fe), manganese (Mn), molybdenum (Mo), selenium (Se), silicon (Si), tin (Sn), vanadium (V), and zinc (Zn).		less than 1.0

FIGURE 7.3 List of essential elements and their percentage.

fluorosis and skeletal fluorosis which is most commonly associated with joint stiffness and joint pain very similar to that of arthritis.[46–48]

In addition to these ions, large quantities of color-containing water are produced by textile dyeing and the dye manufacturing industries. According to incomplete statistics, there are more than 10,000 types of dyes in commercial circulation. Every year, about 7×10^5 tons of dyes is prepared globally and 10–15% is discharged into water bodies as effluents that seriously pollute the environment and affect aquatic organisms. Moreover, they may also harm human health because most dyes are toxic. Dyes are the dangerous class of the pollutants, once enters into water, it becomes difficult to treat, since they are recalcitrant organic molecules which resist to aerobic digestion, and are stable to light, heat, and oxidizing agents.[49–52] Most of the dyes are toxic, whereas some are carcinogenic. Some of the examples of toxic dyes are malachite green, tropaeolin, and CR which are well-known human carcinogens.

Azo dyes is a class of synthetic organic colorants, having azo groups (–N=N–).[53] These dyes offer a wide spectrum of colors and are used for coloring leather, clothes, food, toys, plastics, and cosmetics.

These dyes are relatively resistant to degradation under aerobic conditions; but reduced readily under anaerobic conditions by azoreductase in liver cells, intestinal bacteria, and microflora to form aromatic amines (Fig. 7.4).[54] Therefore, environmental agencies from various countries are now

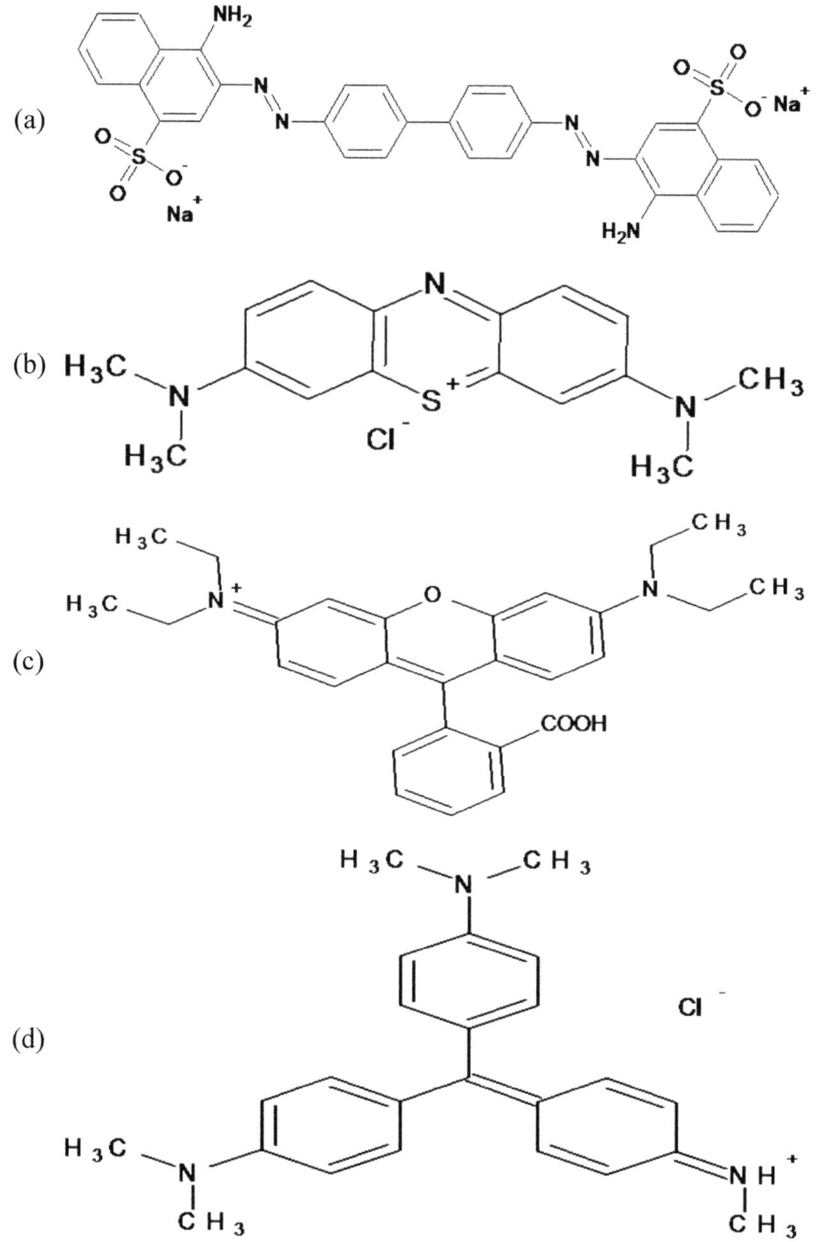

FIGURE 7.4   Various types of dyes structures: (a) methylene blue, (b) Congo red, (c) Rhodamine B, and (d) methyl violet.

concerning to cover up the data of concentration of these contaminants in water bodies and spending a large percentage of their budget to sure the safety of water from their high concentration in order to reduce the environmental and health problems induced by their poisoning.[55] Furthermore, to gain more accuracy in environmental testing of water, environmental agencies gave the maximum limits of contaminants in water to secure the civilization from the effect of high concentration of hazardous contaminants. For example, WHO and USEPA recommended the maximum arsenic limit in drinking water of 0.01 mg/L.[56,57] Similarly, the Ministry of Health, Labour and Welfare, Japan recommended the drinking water quality standards for 46 items to the surety of supply of clean and safe drinking water to civilization.

Therefore, in order to measure for water quality whether drinking water meets the standards perfectly by these measures or not, and to apply these standards, improvement of water treatment facilities, as well as introduction of necessary operational management is required. Simply, these recommended standards can be achieved by the reduce the concentration of contaminant till recommended limits from water through oxidation and precipitation, coagulation, electrocoagulation and coprecipitation, membrane technology, reverse osmosis, electrodialysis, and ion exchange technology.[58–62] However, these treatment technologies require sophisticated instrumentation, high processing cost, and a perfect hand. Besides, removal of pollutants through adsorption is cost-effective, does not involve sophisticated instrumentation, do not require the long procedure, simple in operation and safe to handle can be used ranging from household module to community level.[63,64]

The attraction between the solid surface and pollutants is adsorption process which requires solid particle referred as an adsorbent with a smaller diameter, high surface area, porous and amorphous nature, and having a large number of active sites to attract the pollutants from water. A large number of solid surfaces with above quality have been applied for water purification depicted in various literatures; however, these could not be very effective at high contaminant concentration and even produce the toxic sludge. Among various adsorbent, carbon-based adsorbent having porous and amorphous nature has commonly used materials that showed the better result for water purification.[65]

To ensure the high-performance water purification, lots of advancement in adsorbent have been made, where advanced nanomaterials such as carbon nanotubes (CNTs) and their composite materials are showing promising results.[66] Similar to others, these nanomaterials are also based on common adsorption strategy for removing contaminant species but these materials have also shown promising result as size exclusion membrane filters that allow the flow of water but block the flow of contaminants.[67] This makes it a most advanced technique for water purification; moreover, further chemical modifications in CNTs have also been adapted to decay of contaminant as well as neutralization of harmful waterborne pathogens.[68] The current state of performance of recently developed CNTs and their composites for water purification has been explored below.

7.4   CARBON NANOTUBES

From the last two decades, nanomaterials are becoming more and more concerned subject for researchers at the laboratory as well as commercial level applications due to their remarkable characteristics.[69] Nanomaterials can be defined in a different way according to countries and their application but simply can be defined as "A natural, incidental or manufactured material containing particles, in an unbound state or as an aggregate or as an agglomerate and where, for 50% or more of the particles in the number size distribution, one or more external dimensions are in the size range 1–100 nm."[70] Officially, it has been reported that in last 10 years, the application of nanoparticles increased by more than 3000% till 2015.[71]

Among these nanomaterials, carbon-based nanomaterials are most fashionable materials that are used frequently in various applications. These carbon-based nanomaterials can be characterized as fullerenes (three dimensions < 100 nm), CNTs (two dimensions < 100 nm), and graphene and related materials (one dimension < 100 nm). Among them, CNTs get an extraordinary attention since their discovery.[72,73] CNTs are nanosized rolled graphene sheets cylindrical structure that can be grouped into single-wall CNTs (SWCNTs), double-wall CNTs (DWCNTs) with two concentric tubes, and multiwall CNTs (MWCNTs) with more than two concentric tubes from a few nanometers for SWCNTs to several tens of nanometers for MWCNTs with length of a few micrometers. Due to their remarkable optical, electrical, thermal, mechanical, and chemical properties, they are used in wide range of application.[74] Findings of novel applications of CNTs in environmental fields, particularly in the water purification has become more interesting and concerning subject which focused on use of CNTs as nanomembrane filters for filtration of contaminated water. This technique has become a promising alternative to polymeric membranes for water purification.[75,76] Like activated carbon, CNTs can also be chemically and biologically modified. The lower size of CNTs provides high surface area as well as a high proportion of accessible micropores in their aggregated form. This makes them more suitable for removal of various types of water pollutants giving promising results. Moreover, the chemical modifications of CNTs also improve their quality and selectivity for water purification.[77]

For example, zerovalent iron functionalized MWCNTs was used for the removal of both As(III) and As(V) over the wide range of pH values.[78] The zero-valent iron (ZVI) nanoparticles were functionalized onto oxidized MWCNT using EDTA as chelating agent, ZVI/MWCN/EDTA hybrid (ZCNT) which led to attachment of various functional groups including hydroxyl, amino, and carboxyl groups on their surface; thus, resultant adsorption capacity for As(III) and As(V) was high as 111.1 ± 4.8 and 166.7 ± 5.8 mg/g, respectively.

Similarly, single-pot solid-phase approach was adopted for preparation of magnetic MWCNTs/iron oxide composites, which exhibited good removal capacities for arsenic (47.41 and 24.05 mg/g for As(V) and As(III), respectively) due to their high specific surface area, dispersibility, and magnetic properties.[79] Goethite impregnated graphene oxide (GO)-CNTs aerogel (α-FeOOH@GCA), was used to remove three species of arsenic, As(V), DMA, and p-ASA, and showed large removal efficiency as 56.43, 24.43, and 102.11 mg/g, respectively.[80] Goethite incorporation in GO-CNTs hinders the aggregation of GO-CNTs as well as controls the growth of α-FeOOH nanoparticles. This modification in CNTs enhances the adsorption capacity of CNTs for arsenic species. Similarly, MWCNTs modified with a tertiary amine, ALIQUAT 336, have been developed and utilized for the As(V) detection from their solution through adsorption.[81] The sorption of As(V) on ALIQUAT 336-MWCNTs was confirmed by the X-ray fluorescence spectrometry (Fig. 7.5).

Other than arsenic, modified CNTs showed excellent results for various heavy metals, inorganic ions, organic molecules, and acids. For example, a new tannin resin with CNTs was developed for Pb(II) removal from water and showed higher than 13.8 mg/g removal capacity.[82] Similarly, polyether sulfone/Pb(II)-imprinted MWCNTs mixed matrix membrane,[83]

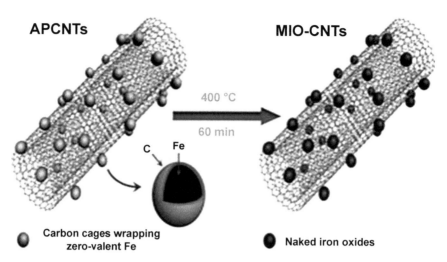

FIGURE 7.5   Adsorption of As(V) on ALIQUAT 336-multiwalled carbon nanotubes (Adapted from Aranda et al.[81] Copyright (2016), with permission from Elsevier).

choline chloride-based deep eutectic solvents functionalized CNTs,[84] and lignin-grafted carbon nanotubes (L-CNTs)[85] were developed and used to remove lead from water. Among them, choline chloride-based deep eutectic solvents functionalized CNTs showed the excellent result for Pb (II) within 15 min at pH 5 with an adsorbent dosage of 5 mg. The maximum adsorption capacity of choline chloride-based deep eutectic solvents functionalized CNTs were obtained (288.4 mg/g). Lignin-grafted CNTs were also used for the removal of oil droplets from water and showed much better result than virgin CNTs (Fig. 7.6).

Similarly, some other heavy metals such as Cd^{2+}, Hg^{2+}, Ni^{2+}, Co^{2+}, and Cu^{2+}, etc. have been separated out from their solution using CNTs. Bhanjana et al.[86] removed Cd^{2+} from their aqueous solutions using CNTs which exhibited 181.8 mg/g of maximum monolayer removal capacity and Ainscough et al.[87] developed epoxidized CNT material for removal of Cd^{2+}, Hg^{2+}, Ni^{2+}, Co^{2+}, and Pb^{2+} from their solution which could be able to remove >99.3% of above-cited pollutants in all cases. Developed material was also capable of microbial removal and showed good antifouling properties. Furthermore, Mohajeri et al.[88] developed the highly ordered CNTs/nanoporous anodic alumina composite membrane for removal of Cu^{2+} and Cd^{2+} from simulated industrial wastewater and Elsehly et al.[89] prepared the

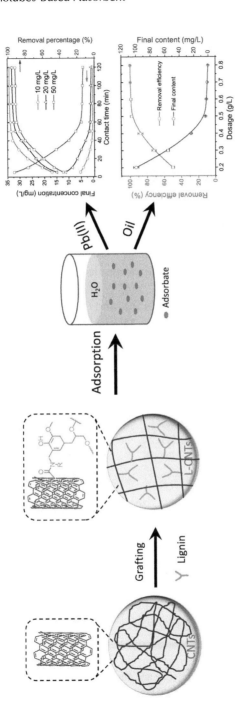

FIGURE 7.6 (See color insert.) Adsorption of Pb (II) on lignin-grafted carbon nanotubes (L-CNTs) (Adapted from Li et al.,[85] Copyright (2017), with permission from Elsevier).

pristine and nitric acid-modified MWCNTs for the filtration of Ni^{2+} ion from their aqueous solution. Ni^{2+} sorption was strongly influenced by pH and could reach 85% at pH = 8.

Moreover, MWCNTs and SWCNTs were also used as an adsorbent for the removal of Cr(VI) ion. Dehghania et al. (Fig. 7.7)[90] used the MWCNTs and SWCNTs for Cr (VI) ion which exhibited the 1.26 and 2.35 mg/g maximum removal capacity, respectively. CNTs are not only capable of heavy metals treatment but are also capable for inorganic ions as well as organic molecules such as dyes, acids, and phenol. Hydroxy-apatite/MWCNTs (HA-MWCNTs) exhibited a 39.22 mg/g of defluoridation capacity and could be able to reduce the fluoride concentration from 8.79 to ~0.25 mg/L in water.[91] MWCNTs were also used for removal of ammonium ion (NH_4^+) from water and showed good result.[92] Radioactive elements such as thorium and uranium were also discarded from aqueous solution using MWCNTs.

Oxidized MWCNTs (ox-MWCNTs) and titanium dioxide (TiO_2) impregnated ox-MWCNTs (TiO_2-ox-MWCNTs) were developed to thorium adsorption in terms of solution pH value and adsorbent doses[93]

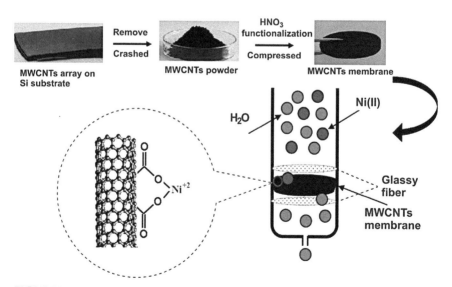

FIGURE 7.7   Adsorption of Ni^{2+} by pristine and nitric acid-modified MWCNTs (Adapted from Elsehly et al.,[89] Copyright (2016), with permission from Elsevier).

ox-MWCNTs and TiO_2-ox-MWCNTs exhibited maximum adsorption capacity of 62.11 and 58.15 mg/g, respectively, for thorium.

Similarly, $CoFe_2O_4$/MWCNTs were developed for uranium (VI) treatment. $CoFe_2O_4$/MWCNTs exhibited fast and efficient sorption for uranium (VI) and showed 212.7 mg/g maximum adsorption capacity at 25°C. $CoFe_2O_4$/MWCNTs could be easily separated from aqueous solutions with a magnet due to their magnetic character (Fig. 7.8).[94]

Organic compounds such as organic dyes, phenol, and humic acid could also be effectively removed from their aqueous solution using MWCNTs.[95,96] For example, MWCNTs/Pd nanotubes exhibited 81.9% of removal capacity for methyl orange after 8 min and 99% after 60 min using 1 mg of dosage per 20 mg/L of the methyl orange.[97] Ponceau 4R (PR), CR, and Allura Red (AR) dyes were also removed from their solution using MWCNT (Fig. 7.9).[98] The amount of adsorbed dye increased in the order PR < AR < CR, and the maximum amount of CR adsorbed onto MWCNT reached of up to 2 μmol m^{-2}.

Similarly, amine functionalized MWCNTs (CNT-NH_2) exhibited exceptional high adsorption capacities, over 500 mg/g, for anionic dyes

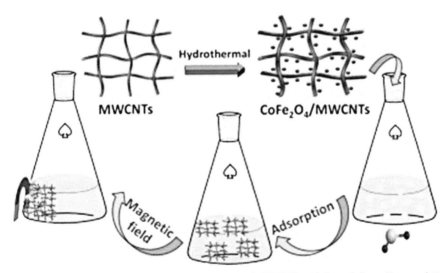

FIGURE 7.8 Adsorption of U(VI) by $CoFe_2O_4$/MWCNTs. (Adapted from Tan et al.[94] Copyright (2015), with permission from Elsevier).

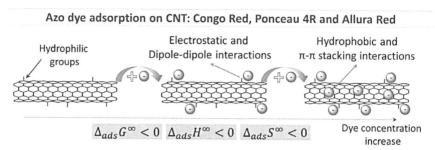

FIGURE 7.9   Adsorption of Ponceau 4R (PR), Congo red (CR) and Allura Red (AR) on MWCNTs (Adapted from Ferreira[98] Copyright (2017), with permission from Elsevier).

(Acid Blue 45 (AB45) and Acid Black 1 (AB1) in single and binary (mixture of dyes) systems.[99] The maximum adsorption capacities of functionalized CNTs were obtained 714 and 666 mg/g, for AB45 and AB1, respectively. These capacities are exceptionally high for the removal of acid dyes. Furthermore, four types of sudan dyes (sudan I, sudan II, sudan III, and sudan IV) were also be separated from their aqueous solution using modified CNTs. For this purpose, magnetic CNTs (MCNTs) have been synthesized by hydrothermal synthesis of Fe_3O_4 nanoparticles onto CNTs.[100] This study investigated that 5 mg MCNTs can remove sudan dyes sufficiently from 20 mL aqueous solution within 60 min at pH in the range of 4–7. MCNTs have also been utilized for the removal of an organic acid such as humic acid (HA). Li et al. developed the magnetic MWCNTs-coated calcium and used it as an adsorbent for the adsorption of HA. Prepared adsorbent showed excellent removal capacity for HA as within 30 min, only 0.5 g/L amount of adsorbent could remove 90.27% of HA from an initial concentration of 20 mg/L at neutral pH (Fig. 7.10).

These are the recently used CNTs for number of water contaminants which exhibited good sorption result for the above-cited pollutant. Therefore, CNTs can be a better option for water purification; however, it needs some more modification.

7.5   SHORTCOMING, FUTURE PROSPECTS, AND CONCLUSION

In this chapter, we have discussed the current status of water pollution and their impact on the water-dependent bodies. Pollutant removal

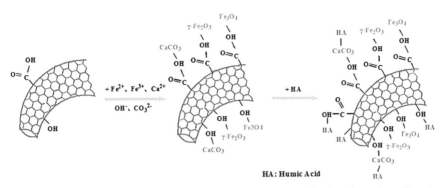

FIGURE 7.10 MWCNTs-coated calcium use as an adsorbent for the adsorption of humic acid (HA) (Adapted from Li et al., Copyright (2017), with permission from Elsevier).

performances of CNTs were discussed thoroughly with their adsorption capacities for various types of pollutants. The current cost of CNTs and their composites is high for water purification which makes them insignificant for commercial as well as domestic use; however, as above-cited CNTs, they may provide the sufficient platform for searching next-generation water purification. Therefore, the concluding remark is that as fabrication cost is decreased, these CNTs will become more suitable and versatile in water purification systems.

ACKNOWLEDGMENTS

The author, Rangnath Ravi is highly thankful to UGC Research Fellowship, UGC Junior Research Fellowship, and Department of Chemistry, Jamia Millia Islamia, New Delhi, India. Dr. Anujit Ghosal is thankful to Government of India, Science and Engineering Research Board (SERB) for financial support in the form of National postdoctoral Fellowship (PDF/2016/003866). The authors are also thankful to the School of Biotechnology, Jawaharlal Nehru University, New Delhi, for implementation of fellowship.

KEYWORDS

- **water contamination**
- **health effect**
- **adsorption**
- **carbon nanotubes**
- **composite**

REFERENCES

1. Herrero, R.; Lodeiro, P.; Rojo, R., et al. The Efficiency of the Red Alga Mastocarpus Stellatus for Remediation of Cadmium Pollution. *Bioresource Technol.* **2008,** *99,* 4138–4146.

2. Chavan, R. P.; Lokhande, R. S.; Rajput, S. I. Monitoring of Organic Pollutants in Thane Creek Water. *Nat. Env. Pollen Technol.* **2005,** *4,* 633–636.

3. Chetna, A.; Pratima, C.; Rina, C. Bacteriological Water Quality Status of River Yamuna in Delhi. *J. Environ. Bio.* **2006,** *27,* 97–101.

4. Bhat, S. C.; Goswami, S.; Ghosh, U. C. Removal of Trace Chromium (VI) from Contaminated Water: Biosorption by Ipomea Aquatica. *J. Environ. Sci. Eng.* **2005,** *47* (4), 316–321.

5. Barik, R. N.; Pradhan, B.; Patel, R. K. Trace Elements in Ground Water of Paradip Area. *J. Indl. Polln. Contl.* **2005,** *21,* 355–362.

6. Pratibha, S.; Subhashini, S.; Shweta, S., et al., Effect of Acute and Chromic Toxicity of Four Commercial Detergents on the Freshwater Fish Gambusia Affinis Bairds Gerard. *J. Environ. Sci. Eng.* **2005,** *47,* 119–124.

7. Shailaja, K.; Mary, J.; Esther, C. Heavy Metals in the Ground Waters of Same Areas of Hyderabad. *Nat. Environ. Pollut. Technol.* **2006,** *5,* 447–449.

8. Sawane, A. P.; Puranik, P. G.; Bhate, A. M. Impact of Industrial Pollution on River Irai, District Chandrapur, with Reference to Fluctuation in CO_2 and pH. *J. Aquatic Bio.* **2006,** *21,* 105–110.

9. Singh, S. H.; Dinesh, K.; Singh, R. V. Improvement of Water Quality Through Biological Denitrification. *J. Environ. Sci. Eng.* **2006,** *48,* 57–60.

10. Singh, R.; Yadav, A. Impact of Carpet Dyeing Units Wastewater on Total Chlorophyl Contents and Biomass of Certain Aquatic Macrophytes. *Indian J. Environ. Sci.* **2005,** *9,* 137–139.

11. Singh, V.; Chandel, C. P. S. The Potability of Groundwater in Terms of Water Quality Index (WQ1) of Jaipur City. *Chem. Environ. Res.* **2004,** *13,* 307–314.

12. Singhal, V.; Kumar, A.; Rai, J. P. N. Bioremediation of Pulp and Paper Mill Effluent with Phanerochaete Chrysosporium. *J. Environ. Res.* **2005,** *26* (3), 525–529.

13. Sonaware D. S.; Shrivastava, V. S. Hazardous Metals in Marine Sediments and Water. *Chem. Environ. Res.* **2004**, *13* (3–4), 221–226.

14. Knezović, Z.; Trgo, M.; Sutlović, D. Monitoring Mercury Environment Pollution Through Bioaccumulation in Meconiumm. *Process Saf. Environ. Protect.* **2016**, 101, 2–8.

15. Chen, W-H.; Yang, W-B.; Yuan, C-S.; Yang, J-C. Zhao, Q-L. Fates of Chlorinated Volatile Organic Compounds in Aerobic Biological Treatment Processes: The Effects of Aeration and Sludge Addition. *Chemosphere* **2014**, *103*, 92–98.

16. Zare, M. R.; Kamali, M.; Omidi, Z.; Kapourchali, M. F. Designing and Producing Large-volume Liquid Gamma-ray Standard Sources for Low Radioactive Pollution Measurements of Seawater Samples by Comparison Between Experimental and Simulation Results. *Measurement* **2016**, *90*, 412–417

17. Tiwari, R. K.; Rajak, G. P.; Mondal, M. R. Water Quality Assessment of Ganga River in Bihar Region, India. *J. Environ. Sci. Eng.* **2005**, *47* (4), 326–355.

18. Tiwari, P.; Saxena, P. N. Response of Biotransformating Organs in *Labeo rohita* to Chromium and Nickel in Yamuna Water at Agra. *J. Ecophysio Occp. Health* **2005**, *5* (1–2), 37–40.

19. Sonaware, D. S.; Shrivastava, V. S. Hazardous Metals in Marine Sediments and Water. *Chem. Environ. Res.* **2004**, *13* (3–4), 221–226.

20. Shailaja, K; Johnson, M. E. C. Heavy Metals in the Ground Waters of Same Areas of Hyderabad. *Nat. Env. Pollen Technol.* **2006**, *5* (3), 447–449.

21. Athikesavan, S.; Vincent, S.; Velmurugan, B.; Vasuki, R. Accumulation of Nickel in the Different Tissues of Silver Carp (*Hypophthalmichthys molitrix*). *Env. Eco.* **2006**, *24*, (5) (1), 143–146.

22. Escudero-Lourdes, C. Toxicity Mechanisms of Arsenic That Are Shared with Neurodegenerative Diseases and Cognitive Impairment: Role of Oxidativestress and Inflammatory Responses. *Neuro. Toxicol.* **2016**, *53*, 223–235.

23. Bandyopadhyay, A. Role of Oxidative Stress in Arsenic (III) Induced Genotoxicity in Cells of Meristematic Tissue of Allium Cepa: An In Vivo Study. *Mater. Today Part A* **2016**, *3*, 3194–3199.

24. Ogun, M.; Ozcan, A.; Karaman, M., et al. Oleuropein Ameliorates Arsenic Induced Oxidative Stress in Mice. *J. Trace Elem. Med. Biol.* **2016**, *36*, 1–6.

25. Aposhian, H. V. *Biochemical Toxicology of Arsenic*; Hodgson, E., Bend, J. R., Philpot, R. M., Eds.; 1989; pp 265–299.

26. Tseng, C. H. Arsenic Methylation, Urinary Arsenic Metabolites and Human Diseases: Current Perspective. *J. Environ. Sci. Health C. Environ. Carcinog. Ecotoxicol. Rev.* **2007**, *25*, 1–22.

27. Siddiqui, S. I.; Chaudhry, S. A. Iron Oxide and Its Modified Forms as an Adsorbent for Arsenic Removal: A Comprehensive Recent Advancement. *Process Saf. Environ. Protect.* **2017**, *III*, 592–626.

28. Acien, A. N.; Guallar, E.; Silbergeld, K. E.; Rothenberg, J. S. Lead Exposure and Cardiovascular Disease: A Systematic Review. *Environ. Health Perspect.* **2007**, *115* (3), 472–482.

29. Woodruff, T. J.; Carlson, A.; Schwartz, J. M.; Guidice, L. C. Proceedings of the Summit on Environmental Challenges to Reproductive Health and Fertility: Executive Summary. *Fertil. Steril.* **2008,** *89* (2) 281–300.

30. Zahid, Z. R.; Al-Hakkak, Z. S.; Kadhim, A. H. H. Comparative Effects of Trivalent and Hexavalent Chromium on Spermatogenesis of the Mouse. *Toxicol. Environ. Chem.* **1990,** *25,* 131–136.

31. Zhang, J.; Li, X. Cancer Mortality in a Chinese Population Exposed to Hexavalent Chromium in Water. *J. Occup. Environ. Med.* **1997,** *39* (4), 315–319.

32. Godt, J.; Scheidig, F.; Siestrup, C. G.; Esche, V.; Brandenburg, P.; Reich, A. The Toxicity of Cadmium and Resulting Hazards for Human Health. *J. Occup. Med. Toxicol.* **2006,** *1,* 22.

33. Buchet, J. P.; Lauwerys, R.; Roels, H.; Bernard, A.; Bruaux, P.; Claeys, F. Renal Effects of Cadmium Body Burden of the General Population. *Lancet* **1990,** *336,* 699–702.

34. Kitaura, K.; Chone, Y.; Satake, N.; Akagi, A.; Ohnishi, T.; Suzuki, Y.; Izumi, K. Role of Copper Accumulation in Spontaneous Renal Carcinogenesis in Long-Evans Cinnamon Rats. *Jpn. J. Cancer Res.* **1999,** *90* (4), 385–392.

35. Cavallo, F.; Gerber, M.; Marubini, E.; Richardson, S.; Barbieri, A.; Costa, A.; De Carli, A.; Pujol, H. Zinc and Copper in Breast Cancer: A Joint Study in Northern Italy and Southern France. *Cancer* **1991,** *67* (3), 738–745.

36. Bernard, S. Autism: A Novel Form of Mercury Poisoning. *Med. Hypotheses* **2001,** *56* (4), 462–471.

37. Holmes, A. S. Reduced Levels of Mercury in First Baby Haircuts of Autistic Children. *Int. J. Toxicol.* **2003,** *22,* 277–285.

38. Das, K. K.; Dasgupta, S. Effect of Nickel on Testicular Nucleic Acid Concentrations of Rats on Protein Restriction. *Biol. Trace Elem. Res.* **2000,** *73,* 175–180.

39. Waksvik, H.; Boysen, M. Cytogenic Analysis of Lymphocytes from Workers in a Nickel Refinery. *Mutat. Res.* **1982,** *103,* 185–190.

40. Dixon, H. B. F. The Biochemical Action of Arsenic Acids Especially as Phosphate Analogues. *Adv. Inorg. Chem.* **1997,** *44,* 191–227.

41. Ma, L.; Li, J.; Zhan, Z., et al., Specific Histone Modification Responds to Arsenic-Induced Oxidative Stress. *Toxicol. Appl. Pharmacol.* **2016,** *302,* 52–61.

42. Lin, S.; Cullen, W. R.; Thomas, D. J. Methylarsenicals and Arsinothiols Are Potent Inhibitors of Mouse Liver Thioredoxin Reductase. *Chem. Res. Toxicol.,* **1999,** *12,* 924–930.

43. Zhang, P.; Lee, J.; Kang, G.; Li, Y.; Zhang, Y. Disparity of Nitrate and Nitrite In Vivo in Cancer Villages as Compared to Other Areas in Huai River Basin, China. *Sci. Total Environ.* **2018,** *612,* 966–974.

44. Lu, Y.; Xu, L.; Shu, W.; Zhou, J.; Qian, G. Microbial Mediated Iron Redox Cycling in Fe (Hydr)oxides for Nitrite Removal. *Bioresource Technol.* **2017,** *224,* 34–40.

45. Bedale, W.; Sindelar, J. J.; Milkowski, A. L. Dietary Nitrate and Nitrite: Benefits, Risks, and Evolving Perceptions. *Meat Sci.* **2016,** *120,* 85–92.

46. Dutta, R. K.; Saikia, G.; Das, B.; Bezbaruah, C.; Das, H. B., Dube, S. N. Fluoride Contamination in Groundwater of Central Assam, India. *Asian J. Water Env. Polln.* **2006,** *3* (2), 93–100.

47. Sahu, A.; Vaishnav, M. M. Study of Fluoride in Groundwater Around the BALCO Korba area (India). *J. Env. Sci. Eng.* **2006**, *48* (1), 65–68.
48. HarishBabu K.; Puttaiah, E. T.; Kumara V.; Thirumala, S. Status of Drinking Water Quality in Tarikere Taluk with Special Reference to Fluoride Concentration. *Nat. Env. Pollen Technol.* **2006**, *5* (1), 71–78
49. Grau, P. Textile Industry Wastewater Treatment. *Water Sci. Technol.* **1991**, *24*, 97–103.
50. O'Neil, C.; Hawkes, F. R.; Hawkes, D. W. Colour in Textile Effluents—Sources Measurement, Discharge Consents and Simulation: A Review. *J. Chem. Technol. Biotechnol.* **1999**, *74*, 1009–1018.
51. Georgious, D.; Melidis, P.; Aivasidis, A.; Gimouhopoulos, K. Degradation of Azo-reactive Dyes by Ultraviolet Radiation in the Presence of Hydrogen Peroxide. *Dyes Pigm.* **2002**, *52*, 97–103.
52. Kuo, W. G. Decolorization Dye Wastewater with Fenton's Reagent. *Water Res.* **1992**, *26*, 881–886.
53. Priscila Maria Dellamatrice, Maria Estela Silva-Stenico, Luiz Alberto Beraldo de Moraes, Marli Fátima Fiore, Regina Teresa Rosim Monteiro, Degradation of Textile Dyes by Cyanobacteria. *Braz. J. Microbiol.* **2017**, *48*, 25–31.
54. Khatri, A.; Peerzada, M. H.; Mohsin, M.; White, M. A Review on Developments in Dyeing Cotton Fabrics with Reactive Dyes for Reducing Effluent Pollution. *J. Clean. Prod.* **2015**, *87*, 50–57.
55. Siddiqui, S. I.; Chaudhry, S. A. Removal of Arsenic from Water Through Adsorption onto Metal Oxide-coated Material. *Mater. Res. Found.* **2017**, *15*, 227–276.
56. Ghosal, A.; Shah, J.; Kotnala, R. K.; Ahmad, S. Facile Green Synthesis of Nickel Nanostructures Using Natural Polyol and Morphology Dependent Dye Adsorption Properties. *J. Mater. Chem.* A **2013**, *1*, 12868–12878.
57. Serra, A.; Brillas, E.; Domènech, X.; Peral, J. Treatment of Biorecalcitrant α-Methylphenylglycine Aqueous Solutions with a Solar Photo-Fenton-aerobic Biological Coupling: Biodegradability and Environmental Impact Assessment. *Chem. Eng. J.* **2011**, *172*, 654–664.
58. Chaudhry, S. A.; Zaidi, Z.; Siddiqui, S. I. Isotherm, Kinetic and Thermodynamics of Arsenic Adsorption onto Iron-zirconium Binary Oxide-coated Sand (IZBOCS): Modelling and Process Optimization. *J. Mol. Liq.* **2017**, *229*, 230–240.
59. Liu, L.; Gao, Z. Y.; Su, X. P.; Chen, X.; Jiang, L.; Yao, J. M. Adsorption Removal of Dyes from Single and Binary Solutions Using a Cellulose-based Bioadsorbent. *ACS Sustainable Chem. Eng.* **2015**, *3*, 432–442.
60. Juneja, S.; Madhavan, A. A.; Ghosal, A.; Moulick, R. G.; Bhattacharya, J. Synthesis of Graphenized Au/ZnO Plasmonic Nanocomposites for Simultaneous Sunlight mediated Photo-catalysis and Anti-microbial Activity. *J. Hazard. Mater.* **2017**.
61. Vilar, V. J. P.; Pinho, L. X.; Pintor, A. M. A.; Boaventura, R. A. R. Treatment of Texile Wastewaters by Solar-driven Advanced Oxidation Processes. *Sol. Energy* **2011**, *85*, 1927–1934.
62. Módenes, A. N.; Espinoza-Quiñones, F. R.; Menenti, D. R.; Borba, F. H.; Palácio, S. M.; Colombo, A. Performance Evaluation of a Photo-Fenton Process Applied to Pol-

lutant Removal from Textile Effluents in a Batch System. *J. Environ. Manage.* **2012,** *104,* 1–8.

63. Ghosal, A.; Ahmad, S. High Performance Anti-corrosive Epoxy–Titania Hybrid Nanocomposite Coatings. *New J. Chem.* **2017,** *41* 4599–4610.

64. Mallampati, R.; Xuanjun, L.; Adin, A.; Valiyaveettil, S. Fruit Peels as Efficient Renewable Adsorbents for Removal of Dissolved Heavy Metals and Dyes from Water. *ACS Sustainable Chem. Eng.* **2015,** *3,* 1117–1124.

65. Siddiqui, S. I.; Chaudhry, S. A. Arsenic Removal from Water Using Nanocomposite: A Review. *Curr. Environ. Eng.* **2017,** *4,* 81–102.

66. Das, R.; Ali, M. E.; Hamid, S. B. A.; Ramakrishna, S.; Chowdhury, Z. Z. Carbon Nanotube Membranes for Water Purification: A Bright Future in Water Desalination. *Desalination* **2014,** *336,* 97–109.

67. Ghosal, A.; Tiwari, S.; Mishra, A.; Vashist, A. Rawat, N. K.; Ahmad, S.; Bhattacharya, J. Design and Engineering of Nanogels. In *Nanogels for Biomedical Applications*; RSC Publishing: 2017; pp 9–28, (978-1-78801-048-1).

68. Brady-Estévez, A. S.; Kang, S.; Elimelech, M. A Single-walled-carbon-nanotube Filter for Removal of Viral and Bacterial Pathogens. *Small* **2008,** *4,* 481–484

69. Liné, C.; Larue, C.; Flahaut, E. Carbon Nanotubes: Impacts and Behaviour in the Terrestrial Ecosystem: A Review. *Carbon* **2017.** DOI: 10.1016/j.carbon.2017.07.089.

70. Buzea, Cristina, Pacheco Ivan, Robbie, Kevin. Nanomaterials and Nanoparticles: Sources and Toxicity. *Biointerphases* **2007,** *2* (4): 17–71.

71. Vance, M. E.; Kuiken, T.; Vejerano, E. P.; McGinnis, S. P.; Jr, M. F. H.; Rejeski, D.; Hull, M. S. Nanotechnology in the Real World: Redeveloping the Nanomaterial Consumer Products Inventory. *Beilstein J. Nanotechnol.* **2015,** *6,* 1769–1780. DOI:10.3762/bjnano.6.181.

72. Arti, V.; Ajeet, K.; Atul, V.; Vidya, S.; Anujit, G. G. Y. K.; Sharif, A.; Madhavan, N. Advances in Carbon Nanotubes–Hydrogel Hybrids in Nanomedicine for Therapeutics. *Adv. Healthcare Mater.* **2018,** *7,* 1701213.

73. Terrones, M. Carbon Nanotubes: Synthesis and Properties, Electronic Devices and Other Emerging Applications. *Int. Mater. Rev.* **2004,** *49,* 325–377.

74. Xu, L.; Wang, Z.; Ye, S. Sui, X. Removal of P-Chlorophenol from Aqueous Solutions by Carbon Nanotube Hybrid Polymer Adsorbents. *Chem. Eng. Res. Des.* **2017,** *123,* 76–83.

75. Ghosal, A.; Vashist, A.; Tiwari, S.; Sharmin, E.; Ahmad, S.; Bhattacharya, J. Nanotechnology for Therapeutics. In *Advances in Personalized Nanotherapeutics*; Springer: 2017, pp 25–40, (978-3-319-63633-7).

76. Park, S. J.; Lee, D. G. Development of Cnt-Metal-Filters by Direct Growth of Carbon Nanotubes. *Curr. Appl. Phy.* **2006,** *6,* 182–186.

77. Sankararamakrishnan, N.; Gupta, A.; Vidyarthi, S. R. Enhanced Arsenic Removal at Neutral pH Using Functionalized Multiwalled Carbon Nanotubes. *J. Environ. Chem. Eng.* **2014,** *2,* 802–810.

78. Chena, B.; Zhua, Z.; Maa, J., et al. One-pot, Solid-phase Synthesis of Magnetic Multiwalled Carbon Nanotube/Iron Oxide Composites and Their Application in Arsenic Removal. *J. Colloid Interface Sci.* **2014,** *434,* 9–17.

79. Fua, D.; Hea, Z.; Sua, S., et al. Fabrication of α-FeOOH Decorated Graphene Oxide-carbon Nanotubes Aerogel and Its Application in Adsorption of Arsenic Species. *J. Colloid Interface Sci.* **2017,** *505,* 105–114

80. Aranda, P. R.; Llorens, I.; Perino, E.; Vito, I. D.; Raba, J. Removal of Arsenic(V) Ions from Aqueous Media by Adsorption on Multiwall Carbon Nanotubes Thin film Using XRF Technique. *Environ. Nanotech. Monit. Manag.* **2016,** *5,* 21–26.

81. Luzardo, F. H. M.; Velasco, F. G.; Correia, I. K. S.; Silva, P. M. S.; Salay, L. C. Removal of Lead Ions from Water Using a Resin of Mimosa Tannin and Carbon Nanotubes. *Environ. Technol. Innovate.* **2017,** *7,* 219–228.

82. Ghosal, A.; Mishra, A.; Tiwari, S. Polymers and Nanocomposites for Biomedical Applications. In *Biopolymers and Nanocomposites for Biomedical and Pharmaceutical Applications;* Nova Science Publisher: 2017, pp 1–16, (978-1-53610-635-0).

83. Al Omar, M. K.; Alsaadi, M. A. H.; Hayyan, M., et al. Lead Removal from Water by Choline Chloride Based Deep Eutectic Solvents Functionalized Carbon Nanotubes. *J. Mol. Liq.* **2016,** *222,* 883–894.

84. Li, Z.; Chen, J.; Ge, Y. Removal of Lead Ion and Oil Droplet from Aqueous Solution by Lignin-grafted Carbon Nanotubes. *Chem. Eng. J.* **2017,** *308,* 809–817.

85. Bhanjana, G.; Dilbaghia, N.; Kim, K. H.; Kumar, S. Carbon Nanotubes as Sorbent Material for Removal of Cadmium. *J. Mol. Liq.* **2017,** *242,* 966–970.

86. Ainscougha, T. J.; Perry, A.; Darren, L. O-R.; Andrew, R.; Barronac, A. Hybrid Super Hydrophilic Ceramic Membrane and Carbon Nanotube Adsorption Process for Clean Water Production and Heavy Metal Removal and Recovery in Remote Locations. *J. Water Process Eng.* **2017,** *19,* 220–230.

87. Mohajeri, M.; Akbarpour, H.; Khani, V. K. Synthesis of Highly Ordered Carbon Nanotubes/Nanoporous Anodic Alumina Composite Membrane and Potential Application in Heavy Metal Ions Removal from Industrial Wastewater. *Mater. Today: Proc.* **2017,** *4,* 4906–4911.

88. Elsehly, E. M.; Chechenin, N. G.; Makunin, A. V., et al. Characterization of Functionalized Multiwalled Carbon Nanotubes and Application as an Effective Filter for Heavy Metal Removal from Aqueous Solutions. *Chinese J. Chem. Eng.* **2016,** *24,* 1695–1702.

89. Dehghania, M. H.; Taher, M. M.; Bajpai, A. K., et al. Removal of Noxious Cr (VI) Ions Using Single-walled Carbon Nanotubes and Multi-walled Carbon Nanotubes. *Chem. Eng. J.* **2015,** *279,* 344–352.

90. Ruan, Z.; Tian, Y.; Ruan, J., et al. Synthesis of Hydroxyapatite/Multi-walled Carbon Nanotubes for the Removal of Fluoride Ions from Solution. *Appl. Surf. Sci.* **2017,** *412,* 578–590.

91. Moradi, O. Applicability Comparison of Different Models for Ammonium Ion Adsorption by Multi-walled Carbon Nanotubes. *Arabian J. Chem.* **2016,** *9,* S1170–S1176.

92. Yavari, R.; Asadollahi, N.; Mohsen, M. A. Preparation, Characterization and Evaluation of a Hybrid Material Based on Multiwall Carbon Nanotubes and Titanium Dioxide for the Removal of Thorium from Aqueous Solution. *Prog. Nucl. Energy* **2017,** *100,* 183–191.

93. Tan, L.; Liu, Q.; Jing, X., et al. Removal of Uranium (VI) Ions from Aqueous Solution by Magnetic Cobalt Ferrite/Multiwalled Carbon Nanotubes Composites. *Chem. Eng. J.* **2015**, *273*, 307–315.

94. Yuab, F.; Lia, Y.; Hana, S.; Mab, J. Adsorptive Removal of Antibiotics from Aqueous Solution Using Carbon Materials. *Chemosphere* **2016**, *153*, 365–385.

95. Asmaly, H. A.; Abussaud, B. Ihsanullah, et al. Ferric Oxide Nanoparticles Decorated Carbon Nanotubes and Carbon Nanofibers: From Synthesis to Enhanced Removal of Phenol. *J. Saudi Chem. Soc.* **2015**, *19*, 511–520.

96. Canoa, O. A.; González, C. A. R.; Paz, J. F. H., et al. Catalytic Activity of Palladium Nanocubes/Multiwalled Carbon Nanotubes Structures for Methyl Orange Dye Removal. *Catal. Today* **2017**, *282*, 168–173.

97. Ferreira, G. M. D.; Ferreira, G. M. D.; Hespanhol, M. C., et al., Adsorption of Red Azo Dyes on Multi-walled Carbon Nanotubes and Activated Carbon: A Thermodynamic Study. *Colloids Surf. A: Physicochem. Eng. Asp.* **2017**, *529*, 531–540.

98. Maleki, A.; Hamesadeghi, U.; Daraei, H., et al. Amine Functionalized Multi-walled Carbon Nanotubes: Single and Binary Systems for High Capacity Dye Removal. *Chem. Eng. J.* **2017**, *313*, 826–835.

99. Sun, X.; Chang, H. O.; Miao, F.; Chen, L. Removal of Sudan Dyes from Aqueous Solution by Magnetic Carbon Nanotubes: Equilibrium, Kinetic and Thermodynamic Studies. *J. Indus. Eng. Chem.* **2015**, *22*, 373–377.

100. Li, S.; He, M.; Li, Z.; Li, D.; Pan, Z. Removal of Humic Acid from Aqueous Solution by Magnetic Multi-walled Carbon Nanotubes Decorated with Calcium. *J. Mol. Liq.* **2017**, *230*, 520–528.

CHAPTER 8

POLYLACTIC ACID/CARBON NANOTUBES-BASED NANOCOMPOSITES FOR BIOMEDICAL APPLICATIONS

FAHMINA ZAFAR*[1], ERAM SHARMIN[2], HINA ZAFAR[3], and NAHID NISHAT[1]

[1]*Inorganic Materials Research Laboratory, Department of Chemistry, Jamia Millia Islamia, New Delhi, India*

[2]*Department of Pharmaceutical Chemistry, College of Pharmacy, Umm Al-Qura University, Makkah Al-Mukarramah, Saudi Arabia*

[3]*Department of Chemistry, Aligarh Muslim University, Aligarh, Uttar Pradesh, India*

Corresponding author. E-mail: fahmzafar@gmail.com

ABSTRACT

Polylactic acid (PLA) is considered as a biodegradable, thermoplastic, and environmentally benign green polymeric material. It has wide applications in biomedical field due to its characteristic features. However, its application is often restricted for particular biomedical applications due to its brittleness, mechanical strength, conductivity, and processibility. These properties can be overcome by the reinforcement of small amount of carbon nanotubes (CNTs) in PLA matrix. In the present chapter, we will discuss briefly about PLA, PLA nanocomposites, CNTs, PLA/CNTs nanocomposites, and their biomedical applications.

8.1   INTRODUCTION

Nowadays biodegradable materials have attracted great attention in versatile applications due to concerns towards environment protection. They can be naturally occurring such as polysaccharides, proteins, and polyesters produced by microorganisms or can be synthesized from biological and renewable sources.[1] Most of the biodegradable polymers are derived from renewable agro-resources such as starch, cellulose, chitin, pectin, PLA, and others. These agro-resources are labeled as environmentally benign materials due to their degradable products that are nonhazardous to the environment.[2,3] Till date, applications of biodegradable polymers have grown to take account of mainly agricultural, biomedical, and food packaging applications.

Among all biodegradable agro-resources, PLA has attracted great attention over the last several decades as important biodegradable polymer for packaging, biomedical, and tissue engineering field among other biomedical applications.[4] This is due to its characteristic features such as high transparency, ease of processing, good mechanical properties, renewable origin, bioresorbability, biodegradability, and biocompatibility with the living organisms together with human body.[5] The aforementioned properties and maneuver degradation rate of PLA have permitted it to engulf the limitation of brittleness along with difficulties for processing of traditional bioactive or bioabsorbable ceramics scaffold in bone and cartilage making.[6] The fabrication of porous PLA-derived scaffold has been carried out by many technologies such as solvent casting, membrane lamination, particular leaching, melt molding, and precise excursion. Along with them, elcetrospinning technique has also been used to process PLA in highly porous ultrafine fibrillar tissue scaffold with 3D structure that is very much mimicking natural extracellular matrix. The drawbacks associated with electrospum PLA scaffolds such as poor mechanical properties and low hydrophilicity limit its utilization in biomedical fields.[6] These drawbacks could be overcome by improving their performance through copolymerization, blending, filling or nanocomposite formation techniques.[2,7-10] Amongst them, the latter technique involving the incorporation of small quantity of nanofiller is a simple technique to improve the properties of PLA for biomedical applications.[3,11]

The reinforcement of nanofillers in very little amount (<5 wt%) into the polymer matrix results in remarkable improvement of its performance as compared to those of virgin polymers or conventional micro-/macro-composites. The enhanced performances such as mechanical strength, elastic modulus and dimensional stability, chemical resistance, gas, water, and hydrocarbon permeability, thermal stability, heat distortion temperature, physical weight, and electrical conductivity along with biodegradability have been reported for green nanocomposites.[12] The enhancements of these properties would be correlated to the nanoscale fillers that have large specific surface area and their strong electrostatic interactions with polymers. The number of nanoparticles including fibers,[5] organically modified layered silicates nanoclay[2], fibrous silicates,[13] polysaccharide nanowhiskers,[12,14] and carbon nanotubes (CNTs)[15] have been used to improve the performance of PLA for biomedical applications.[9] The goal of this chapter is to describe and discuss the state-of-the-art for the production of PLA/CNT nanocomposites intended for biomedical applications.

8.2 POLYLACTIC ACID NANOCOMPOSITES

Polylactic acid (PLA) is biodegradable, thermoplastic, linear, and aliphatic polyester with either a semicrystalline or amorphous structure. It can be produced from renewable resource derived monomers such as 2-hydroxypropionic acid (lactic acid, LA) via the renewable sources such as carbohydrate feedstocks (maize, wheat, whey, corn, potato, or molasses) fermentation followed through either by condensation or ring-opening polymerization (Fig. 8.1). In general, industrial PLA grade are copolymers of poly(L-LA) and poly(D, L-LA) that are produced respectively, from L-lactides and D, L-lectides. PLA properties depend on ratio of L-enantiomers and D, L-enantiomers. The property of PLA depends on the ratio of such L-enantiomers to D, L-enantiomers that also affect the semicrystalline and amorphous behavior. Polymers derived from L-LA are highly crystalline in nature with very high melting points, while mixture of D- and L-LA results in the amorphous polymers with a low glass transition temperature (Tg).

Due to biodegradability, biocompatibility, and good mechanical strength of PLA and its derivatives such as copolymers, they have been used in bio-assimilable sutures and drug-loading devices but due to its high cost, brittleness, slow crystallization, and poor barrier properties, application aspects are restricted as compared to nondegradable materials. Several modifications have been carried out to overcome these drawbacks.[2,16,17]

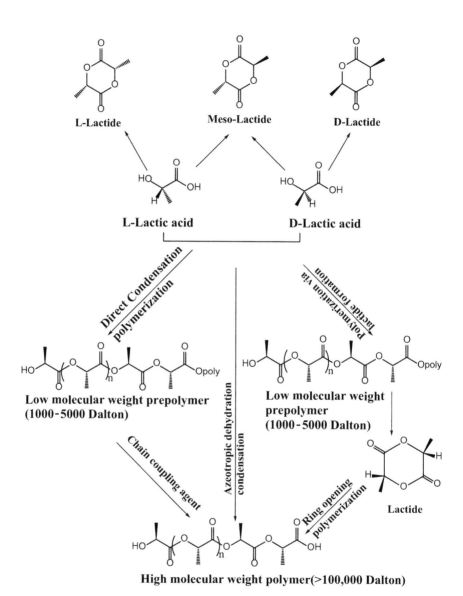

FIGURE 8.1 Stereoisomers of lactic acid, lactide, and polymerization route to prepare PLA.

The properties of PLA are enhanced for its use in biomedical applications by the reinforcement of nanofillers in PLA. There are different nanofillers such as clay, polysaccharide whiskers, and carbon nanotubes (CNTs) that have been used to enhance the performance of PLA.[5,9,12] Generally, reinforcement of these fillers into PLA to form PLA nanocomposites can be carried out by the following methods[18]:

i)   In situ polymerization[19]
ii) Template method
iii) Solvent mixing[20]
iv) Melt mixing[21]
v)   Solid state mixing
vi) Sol–gel

8.3   PLA/CNTs NANOCOMPOSITES

CNTs has been used as conductive fillers that are associated with characteristic features such as excellent flexibility, low mass density, high aspect ratio (characteristically > 1000), and very high strength along with tensile moduli. It can be divided into single-walled CNTs (SWCNTs) or double-walled CNTs (DWCNTs) or multiwalled CNTs (MWCNTs) on the basis of the number of tubular walls that are rolled up from a graphene sheet. Individual SWCNTs (~0.7–1.5 nm diameter ranges) can be metallic or semiconducting. MWCNTs (innermost tubular layer diameters ranges ~2–10 nm) have distinctive features such as 1D nanostructure along with low density and high aspect ratio. (Figs. 8.2 and 8.3)[22]

They have excellent properties such as mechanical, electrical, and thermal conductivity and have acquired considerable interests in number of biomedical applications.[23] Nowadays, CNTs' functionalization is carried out to enhance their use in biomedical applications, and to avoid problems associated with the same such as potential toxicity, solubility, dispersibility in water, and others. CNTs' functionalization has been carried out via surface modification of CNTs in strong acidic conditions such as concentrated H_2SO_4 and HNO_3. Generally, the functionalization occurred on the surface of CNTs with carboxylic (–COOH), hydroxyl (–OH), thiol (–SH), and amide (–$CONH_2$) and polymers such as PLA, poly(LA-glycolic acid), polyglycolic acids, and different biopolymers or natural polymers (Fig. 8.4).[24]

(a)   (b)

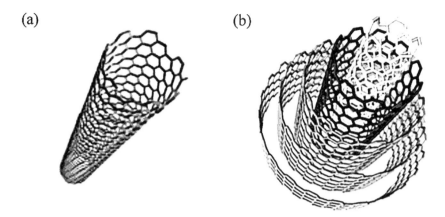

FIGURE 8.2 Schematic representation of (a) SWNT and (b) MWNT (Russian Doll model). (Reproduced with permission from Ref. [9]. © 2013 Elsevier.)

(a)   (b)

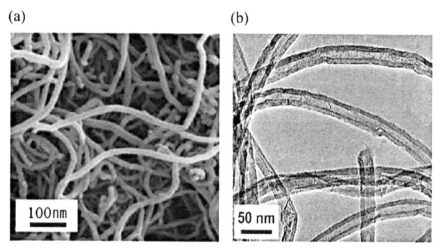

FIGURE 8.3 SEM (a) and TEM (b) images of MWCNT. (Reproduced with permission from Ref. [22]. © 2016 Elsevier.)

CNTs-reinforced polymer nanocomposites have gained much attention in comparison to virgin polymers. The potentiality of CNTs provided their necessary structural reinforcement for biomedical applications such as scaffold tissue engineering, bone implants, drug delivery, biosensors, and others.[9,15] A significant improvement in mechanical, thermal, and electrical properties of the composites has been observed with the dispersion of very small fraction of CNTs into a PLA polymer.[25] These enhancements depend not only on the excellent properties of CNTs but also on their adhesion, alignment, and dispersion in polymer matrix along

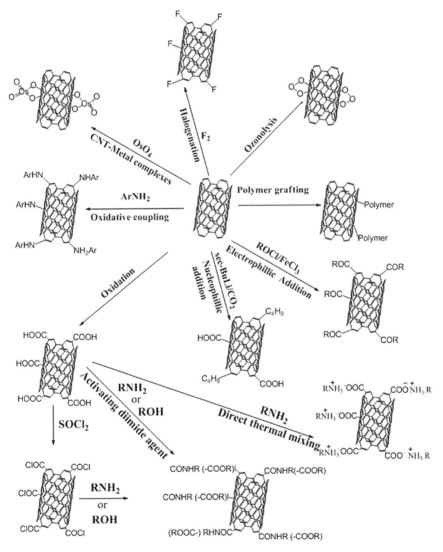

FIGURE 8.4 Functionalization of CNTs and derivatization of CNT–COOH with amine and alcohol.

with the degree of surface modification. In fabrication of nanocomposites, CNTs aggregate due to curling and twisting that dramatically hamper the properties of the final nanocomposites. The excellent properties (very high strength and stiffness) of CNTs would not be completely transferred into the final materials. So, individual CNTs within a polymer only exhibit a fraction of such excellent properties. For their better or homogenous dispersion and alignment, several chemical modifications have been done quite successfully.[11]

The biodegradable, biocompatible polymer poly(L-lactide) reinforced with conducting CNTs are generally prepared through solution blending, precipitation, melt mixing, and in situ polymerization methods while PLA/CNTs fibers were developed by solution blow spinning methods (Fig. 8.5).[15,28] The resulting nanocomposite is formed via more hydrophobic C–CH_3 groups interaction as suggested by Raman spectroscopy. The direct current (DC) electrical transport property of the composite with loading of MWCNTs was described by percolation mechanism. The discrepancy in the values of conductivity was observed due to different levels of MWCNTs dispersion along with the probable aspect ratios difference of CNTs.[25] MWCNTs lower down the Tg, crystallization temperature (Tc), and melting temperature (Tm) along with the formation of new crystallites of the final nanocomposites as compared to virgin PLA. These results corroborate that CNTs have the effect of plasticizing PLA matrix. This effect can be determined by differential scanning calorimetry (DSC). Field emission scanning electron microscopy (FE SEM) and high-resolution transmission electron microscopy (HR TEM) of PLA/MWCNT composites reveals that PLA absorb on the surface of MWCNTs (Fig. 8.6).[29] The mechanical tensile strength of the nanocomposites was significantly improved with very small amount of modified CNTs (mCNTs)–calcium phosphate (CP) hybrid nanopowders.[30] The enhancement of tensile strength used for bone tissue engineering or bone nails applications. The dispersion of small quantity of carboxylic group-modified MWCNTs (c-MWCNTs) into PLA via in situ polymerization increases the Tg and electrical conductivity, whereas decreases thermal decomposition temperature of resulting PLA/c-MWCNTs composite. This is due to an even dispersion of the same in PLA matrix and PLA-coated on the c-MWCNTs' surface.[19] The polyurethane (PU)/PLA (90:10 wt% of PU:PLA) blend reinforced with UV/ozone-modified MWCNTs (2–10 wt%) were developed via melt mixing process. The surface-modified CNTs significantly

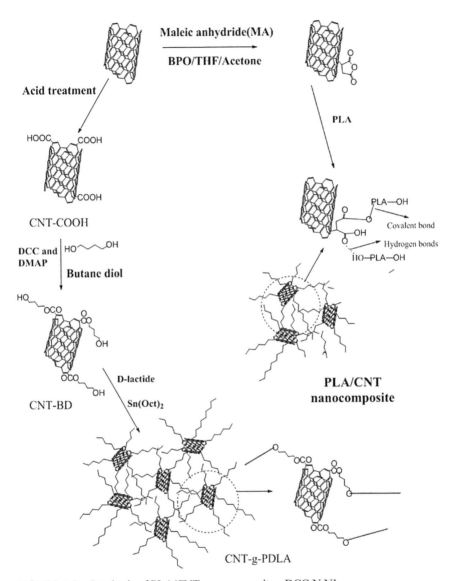

FIGURE 8.5 Synthesis of PLA/CNT nanocomposites. DCC:N,N'-dicyclohexylcarbodiimide and DMPA: N,N-dimethylpropionamide.[26,27]

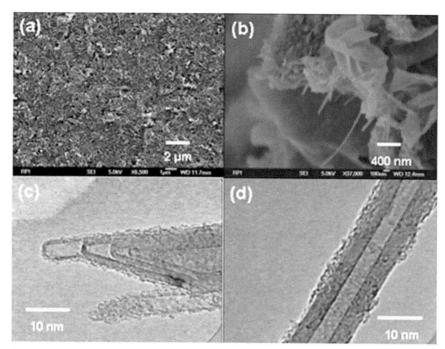

FIGURE 8.6   FE SEM images of PLA/MWCNT (3 wt% MWCNT) (a), magnified FE SEM image of PLA/MWCNT (3 wt% MWCNT) (b), HR TEM of PLA/MWCNT (7 wt% MWCNT) showing PLA absorb on surface of MWCNTs (c) and (d). (Reproduced with permission from Ref. [29]. © 2006 American Chemical Society.)

enhanced tensile strength, dynamic storage modulus, and Tg as compared to pristine CNT-loaded polymer blends. These enhancements of the properties were corroborated to higher polymer-CNT interactions. The excellent electrical and thermal conductivity of PU/PLA-modified CNTs than pristine CNT-loaded system with same concentration were due to the fine dispersion in the former system.[21]

8.4   BIOMEDICAL APPLICATIONS OF PLA/CNT NANOCOMPOSITES

PLA/CNT nanocomposites have been investigated for variety of biomedical applications (Table 8.1).[31] They have been used as new generation implant materials which would result in stimulation of cell growth along with tissue generation by facilitating the physicoelectrical signal transfer.[32] The inhibition of 10T1/2 fibroblast cells growth in the presence of MWCNT (7 wt%) in the PLA/MWCNT nanocomposites at different incubation intervals that can be due to the less favorable attachment of cells to the surface of nanocomposite (Fig. 8.7).[29]

PLA combined with mCNTs-CP can be used for dentistry, orthopedics, and maxillofacial surgery application as an alternative to metallic implants (such as screws, pins, and plates) for use as fixatives. Moreover, the bionanocomposites with low loading of mCNTs (0.1%) revealed in vitro significantly stimulated biological responses including cell proliferation and osteoblastic differentiation in terms of gene and protein expressions. When very small concentration of CNTs (0.05 wt%) are present in nanocomposites, the cell proliferation and osteogenic differentiation were significantly improved while the mechanical strength was preserved. The bionanocomposite may be considered as a new material of choice for

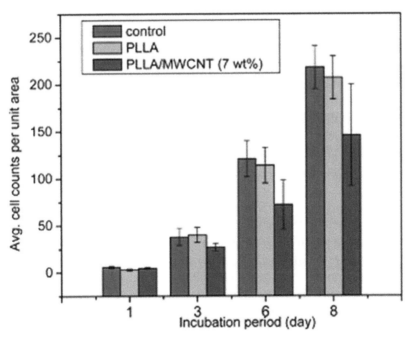

FIGURE 8.7 (See color insert.) Fibroblast cell counts/unit area for control, PLA (PLLA) and PLA/MWCNT (PLLA/MWCNT) (7 wt%). (Reproduced with permission from Ref. [29]. © 2006 American Chemical Society.)

developing hard tissue implants based on above studies.[30] The improvement of electrical conductivity of PLA (with an optimal content of 0.5% for MWCNTs) will resolve problems of high dielectric constant and electrostatic problem for PLA materials.[19] PLA is an insulator and not appropriate for electrically stimulating bone growth; the blend of PLA/CNT is used to expose cells to electrical stimulation. The homogenous, smooth, and nanoporous PLA/CNT (80%/20%w/w) composite was exhibited ideal electrical conductivity for bone growth. The cultured osteoblasts (the bone-forming cells) on the surface of the composite were exposed to electric stimulation (10 μA at 10 Hz, 6 h/day) in vitro. The composite promoted a 46% increase in osteoblasts proliferation and a 307% increase in extracellular calcium for, respectively, 2 and 21 consecutive days, and upregulation of mRNA expression for collagen type-I, for both 1 and 21 consecutive days as compared to virgin PLA. These results corroborated that electrical stimulation is delivered through conducting PLA/CNT nanophase composites, which promotes osteoblast functions (responsible for the chemical composition of the organic and inorganic phases of bone), and the mechanism involved in the new bone formation is at cellular or molecular level.[32] The electro-active shape memory behavior of PU/PLA nanocomposites with surface-modified CNTs exhibit a remarkable recoverability of its shape at lower applied DC voltage as compared to pure/pristine CNTs-loaded system.[21] Electrospun poly[(D,L-Lactic)-co-(glycolic acid)], PLGA, and plasma-functionalized (amine functionalized) SWCNTs (a-SWCNTs) nanocomposites were developed for biomedical applications. The diameter and surface energy of electrospun PLGA/a-SWCNT nanocomposites decreases and increases, respectively, as compared to those of electrospun PLGA/c-SWCNT nanocomposites. The tensile modulus of electrospun PLGA/c-SWCNT and PLGA/a-SWCNT nanocomposites increases, respectively, by 127% and 226% as compared to that of electrospun PLGA membranes that is due to well dispersion and improved adhesion of SWCNTs to the surrounding polymer matrix.[33]

Electrospinning PLA/MWCNT fibers developed for a scaffold for tissue engineering. The diameter of fibers was drastically reduced by 70% with MWCNT to form fibers with a mean diameter of 700 nm that is due to an increased surface charge density for the MWCNT/polymer solution. With 0.25 wt% loading of MWCNT, increases the conductance of the scaffold and the tensile modulus of the composite. An adipose-derived human mesenchymal cells (hMSCs) growth and longitudinal alignment observed after two weeks.[34,35] Randomly (R) and aligned (A) PLA/MCNTs (0–5 wt%) nanofibers developed by solution mixing followed by electrospinning for bone tissue engineering. The mechanical properties such as tensile, young modulus, porosity, and conductivity increase up to 3 wt% MWCNTs, whereas surface resistance decreases with MWCNTs. The electrical conductivity is better in case of aligned fibers as compared to randomly oriented fibers meshes with same MWCNTs contents. PLA/MWCNTs nanofibers degraded faster as compared to PLA. The degradation rate increases with the MWCNTs over incubation. The rate of degradation was found little higher in aligned nanofiber meshes as compared to randomly oriented fiber meshes with same MWCNTs contents. Cell metabolic activity (CellMA) for osteoblasts cells increases 20% and 40%, respectively, for R3 and A3 fibers at day 3 under 100 μA DC as compared to unstimulated control. It was found that the induced osteoblasts alignment 190% for R3 and 87% for A3 fibers meshes at day 3 under 100 μA DC (Fig. 8.8). The aspect ratio for A3 is higher than R3 meshes. These features collectively play an important role in bone tissue engineering (R3 and A3: numeral indicate % of MWCNTs).[28] Highly oriented PLA/MWCNTs (5–15 wt%) nanocomposite was fabricated for their use as blood-contacting medical devices. The mechanical properties (tensile and modulus) and biocompatibility improve with MWCNT and orientation. The oriented PLA and PLA/MWCNTs composite possess desirable blood compatibility, hematolysis rate 0% at 480% Drawing rate (DR) in both cases (permissible rate is <5% according to ASTM F 756-00 for biomaterials). The initial clotting time (time at which OD values equal to 0.1) at 480% DR increase around 37 min as compared to that of isotropic samples and neat PLA. The blood compatibility of the materials is determined by the platelet adhesion and activation on the biomaterial surface. Lower the values of these two parameters denote good blood compatibility. These values decrease with MWNCTs and drawing. At 480% DR, PLA/MWCNTs with 5 wt% MWCNTs sample show prone to prevent adhering platelet.[36]

Highly porous PLA/MWCNT nanocomposite fibers have been developed for their utilization as potential biosensors materials especially in glucose sensors. This composite fibers is fabricated on indium tin oxide (ITO) electrodes via solution blow spinning.[7]

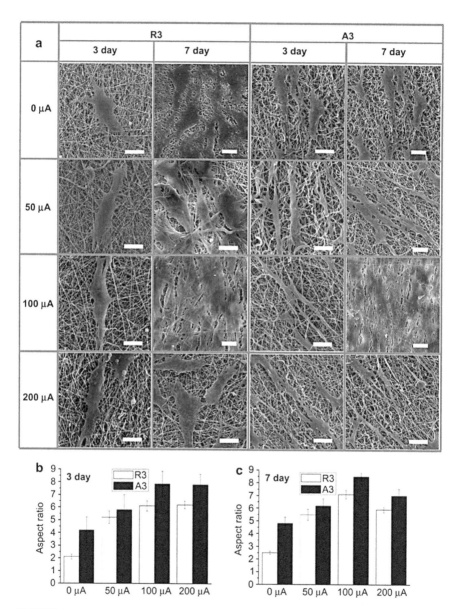

FIGURE 8.8 (a) Selected SEM images of osteoblasts culture on randomly (R3) oriented and aliened (A3) nanofiber meshes with 3 wt% MWCNTs, for 3 and 7 days with different electric stimulation scalebar 30 mm, and (b, c) osteoblasts elongation measured by the aspect ratio after cultured for 3 and 7 days. (Reproduced with permission from Ref. [28]. © 2011 Elsevier.)

TABLE 8.1 Properties of PLA, CNTs, Nanocomposite Composition, Method of Preparation, and Their Applications.

Sr. no.	PLA	CNTs	PLA/CNTs composition and method	Application	References
1	Mw = 100,000 g/mol kg/mol	–	PLA/MWCNT (0–20 wt%) Disperse CNTs in emulsion of PLA followed by sonication and vacuum dry	Bone regeneration	[32]
2	–	mMWCNT-CP hybrid (0.1–0.25 wt% CNT in mCNTs-CP)	mCNTs-CP-PLA (up to 50% PLA) By solution mixing	Hard tissue implant	[30]
3	Mw = 100–150 kg/mol	MWCNT	PLA/MWCNT (0–20 wt%) By solution mixing and precipitation	Biocompatible material	[29]
4	Mw = 250,000 g/mol	MWCNT: l = 15–20 nm, d = 5–15 μm	PLA/MWCNT (0–1 wt%) scaffold (electrospun)	Tissue scaffold	[35]
5	PLA = Mw 180 kDa)	MWCNT: (l = 10–20 μm, d = 10–20 nm)	PLA/MWCNT (0–5 wt%) random (R) and aligned (A) nanofibers Solution mixing and electrospinning	Bone tissue engineering	[28]
6	PLA: Mw: 1 × 10^5 g/mol	MWCNTs: l = 10–30 μm, d = 20–40 nm 1.63 wt% –OH group and small amount of –COOH groups	PLA/MWCNTs (5–15 wt%) Solid hot drawing	Blood contacting medical devices	[36]
7	PLA:Mn = 12 × 10^4 g/mol Mw/Mn: 1.27	MWCNT: d = 10–15 nm	PLA/MWCNT (0–3 wt%) films By solvent casting method	Primary stem cell culture	[20]
8	Mw: 220,000 g/ mol	MWCNT: d = 9.5 nm; l = 1.5 μm	PU/PLA (90:10)/CNT (2–10 wt%) Melt processing techniques	Electro active shape memory	[21]
9	Mw: 75,000 g/ mol	MWCNT	PLA/MWCNT (0–3 wt%) fibers Solution blow spinning	Glucose biosensor	[37]
10	Biopolymer 4032D	MWCNT: d = 15–20 nm, l = 6.5–5.5 μm	PLA/MWCNT-Ag (0–0.3 wt%) Ag to MWCNT ratio was 1:9 By melt blending	Biocompatible and antibacterial	[38]

CNTs, carbon nanotubes; MWCNT, multiwalled carbon nanotube; PLA, polylactic acid; PU, polyurethane.

PLA/MWCNT fibrous electrode is developed for detection of hydrogen peroxide. Cyclic voltameter was used to monitor the electron transfer kinetics in fiber-modified-ITO electrode. The optimum condition for the sensor was observed 1 min of time for fibers deposition with 1 wt% MWCNT loading into the polymer. The sensors demonstrated sensitivity and detection limit for glucose as 358 nA·mM^{-1} and 0.08 mM, respectively. It exhibited a wide linear range (0–900 μ·mM) that follows typical Michaelis–Menten saturation kinetics, and Michaelis–Menten constant (K_M) values for the optimum immobilized glucose oxidase was 4.3 mM. The fibers of blow spun nanocomposite would be of immense potential in amperometric biosensor application because of high porosity, surface-to-volume ratio along with high permeability of the substrate.[37]

MWCNT do not show any antibacterial activity. Nanosilver doped in MWCNT (MWCNT-Ag) to modified MWCNT for their use in the improvement of mechanical and antibacterial properties of PLA. Reinforcement of MWCNT-Ag improves the properties of PLA/MWCNT-Ag nanocomposite as compared to pristine PLA. Thermal stability, tensile strength, Tg, Tc, degree of crystallinity (Xc), and antibacterial activity of PLA increases with MWCNT-Ag. An enhancement of Tg correlated to that MWCNT-Ag impart hindrance for the molecular chain motions in PLA, while higher the value of Tc and Xc correlated to that MWCNT-Ag act as a nucleating agent in the composite. The tensile strength of PLA/MWCNT-Ag nanocomposite increases as the MWCNT-Ag increases from 0–0.3 phr, due to the uniform dispersion that increased the surface area of the MWCNT-Ag which are used for the bonding with PLA matrix.

The higher antibacterial activity against *Staphylococcus aureus* (an aerobic bacterium generally found in burn wounds) observed at 0.1 phr MWCNT-Ag, whereas the activity decreases with 0.2 and 0.3 contents of MWCNT-Ag. It was found that after 72 h of incubation, the relative growth rate (RGR) of PLA was 258.76%, whereas for PLA with MWCNT-Ag 0.1 phr, 0.2 phr, and 0.3 phr contents were, respectively, 176.2%, 126.39%, and 65.98%. The RGR values decreases with MWCNT-Ag contents due to the release of MWCNT-Ag. Therefore, higher biocompatibility of the composite was found with the MWCNT-Ag contents less than 0.3 phr.[38]

8.5 CONCLUSION

PLA reinforced with nanofillers is utilized in several applications. PLA nanocomposites with functionalized CNTs have found biomedical applications as in cell proliferation and osteogenic differentiation, hard tissue implants, antibacterial, and others. However, the area still needs investigations and extensive research unfolding several known and unknown applications of such nanocomposites.

ACKNOWLEDGMENTS

Dr. Fahmina Zafar is thankful to Department of Science & Technology, New Delhi, India for a project under the Women Scientists Scheme (WOS) for Research in Basic/Applied Sciences (Ref. No. SR/WOS-A/CS-97/2016) and the Head, Department of Chemistry, Jamia Millia Islamia (JMI), for providing facilities to carry out the research work.

KEYWORDS

- **renewable resource**
- **polylactic acid**
- **carbon nanotubes**
- **nanocomposite**
- **biomedical applications**

REFERENCES

1. Zafar, F.; Sharmin, E.; Shreaz, S.; Zafar, H.; Mir, M. U. H.; Behbehani, J. M.; et al. Biobased Pharmaceutical Polymer Nanocomposite: Synthesis, Chemistry and Antifungal Study. In *Handbook of Polymers for Pharmaceutical Technologies: Structure and Chemistry*; Scrivener Publishing LLC: Beverly, MA; 2015; Vol. 1, pp 327–350.
2. Liu, M.; Pu, M.; Ma, H. Preparation, Structure and Thermal Properties of Polylactide/Sepiolite Nanocomposites with and Without Organic Modifiers. *Compos. Sci. Technol.* **2012**, *72*, 1508–1514.

3. Ghosal, A.; Mishra, A.; Tiwari S. Polymers and Nanocomposites for Biomedical Applications. In *Biopolymers and Nanocomposites for Biomedical and Pharmaceutical Applications*. Nova Science Publishers, Inc.: New York, 2017; Chapter 1, pp 1–16.

4. Chevalier, M. T.; Martin-Saldaña, S.; Alvarez, V. A. Poly (Lactic-co-glycolic Acid) Nanoparticles: Current Advances in Tumor and Metastasis Targeted Therapies. In *Biopolymers and Nanocomposites for Biomedical and Pharmaceutical Applications*; Nova Science Publishers, Inc., New York, 2017, pp 93-106.

5. Armentano, I.; Bitinis, N.; Fortunati, E.; Mattioli, S.; Rescignano, N.; Verdejo, R.; et al. Multifunctional Nanostructured PLA Materials for Packaging and Tissue Engineering. *Prog. Polym. Sci.* **2013,** *38,* 1720–1747.

6. Zhou, C.; Shi, Q.; Guo, W.; Terrell, L.; Qureshi, A. T.; Hayes, D. J.; et al. Electrospun Bio-nanocomposite Scaffolds for Bone Tissue Engineering by Cellulose Nanocrystals Reinforcing Maleic Anhydride Grafted PLA. *ACS Appl. Mater. Interfaces.* **2013,** *5,* 3847–3854.

7. Oliveira, J. E.; Zucolotto, V.; Mattoso, L. H.; Medeiros, E. S. Multi-walled Carbon Nanotubes and Poly (Lactic Acid) Nanocomposite Fibrous Membranes Prepared by Solution Blow Spinning. *J. Nanosci. Nanotechnol.* **2012,** *12,* 2733–2741.

8. Park, S.; Abdal-Hay, A.; Lim, J. Biodegradable Poly (Lactic Acid)/Multiwalled Carbon Nanotube Nanocomposite Fabrication Using Casting and Hot Press Techniques. *Arch. Metall. Mater.* **2015,** *60,* 1557–1159.

9. Raquez, J-M.; Habibi, Y.; Murariu, M.; Dubois, P. Polylactide (PLA)-based Nanocomposites. *Prog. Polym. Sci.* **2013,** *38,* 1504–1542.

10. López-Rodríguez, N.; López-Arraiza, A.; Meaurio, E.; Sarasua, J. Crystallization, Morphology, and Mechanical Behavior of Polylactide/Poly (ε-caprolactone) Blends. *Polym. Eng. Sci.* **2006,** *46,* 1299–1308.

11. Mina, M. F.; Beg, M. D.; Islam, M. R.; Nizam, A. K. A. A.; Younus, R. M. Characterization of Biodegradable Nanocomposites with Poly (Lactic Acid) and Multi-walled Carbon Nanotubes. World Academy of Science, Engineering and Technology, *Int. J. Mater. Metall. Eng.* **2013,** *7,* 74–79.

12. Sinha, A.; Martin, E. M.; Lim, K-T.; Carrier, D. J.; Han, H.; Zharov, V. P.; et al. Cellulose Nanocrystals as Advanced "Green" Materials for Biological and Biomedical Engineering. *J. Biosyst. Eng.* **2015,** *40,* 373–393.

13. Wu, J.; Zou, X.; Jing, B.; Dai, W. Effect of Sepiolite on the Crystallization Behavior of Biodegradable Poly (Lactic Acid) as an Efficient Nucleating Agent. *Polym. Eng. Sci.* **2015,** *55,* 1104–1112.

14. Haafiz, M. M.; Hassan, A.; Khalil, H. A.; Khan, I.; Inuwa, I.; Islam, M. S.; et al. Bionanocomposite Based on Cellulose Nanowhisker from Oil Palm Biomass-filled Poly (Lactic Acid). *Polym. Test.* **2015,** *48,* 133–139.

15. Akbari, A.; Majumder, M.; Tehrani, A. Polylactic Acid (PLA) Carbon Nanotube Nanocomposites. In *Handbook of Polymer Nanocomposites Processing, Performance and Application*; Springer: Berlin, Heidelberg: 2015; pp 283–297.

16. Abayasinghe, N. K.; Perera, K. P. U.; Thomas, C.; Daly, A.; Suresh, S.; Burg, K.; et al. Amido-modified Polylactide for Potential Tissue Engineering Applications. *J. Biomater. Sci. Polym. Ed.* **2004,** 15, 595–606.

17. Jain, R. A. The Manufacturing Techniques of Various Drug Loaded Biodegradable Poly (Lactide-co-glycolide)(PLGA) Devices. *Biomaterials* **2000**, *21*, 2475–2490.

18. Thakur, V. K.; Thakur, M. K.; Kessler, M. R. *Handbook of Composites from Renewable Materials, Biodegradable Materials*; John Wiley & Sons, Scrivener Publishing LLC: Beverly, MA, 2017; Vol 5; pp 1-647

19. Li, Q-h.; Zhou, Q-h., Dan, D.; Yu, Q-z., Li, G.; Gong, K-d.; et al. Enhanced Thermal and Electrical Properties of Poly (D, L-lactide)/Multi-walled Carbon Nanotubes Composites by In-situ Polymerization. *Trans. Nonferrous Met. Soc. China* **2013**, *23*, 1421–1427.

20. Lizundia, E.; Sarasua, J. R.; D'Angelo, F.; Orlacchio, A.; Martino, S.; Kenny, J. M.; et al. Biocompatible Poly (L-lactide)/MWCNT Nanocomposites: Morphological Characterization, Electrical Properties, and Stem Cell Interaction. *Macromol. Biosci.* **2012**, *12*, 870–781.

21. Raja, M.; Ryu, S. H.; Shanmugharaj, A. Thermal, Mechanical and Electroactive Shape Memory Properties of Polyurethane (PU)/Poly (Lactic Acid)(PLA)/CNT Nanocomposites. *Eur. Polym. J.* **2013**, *49*, 3492–3500.

22. Jang, I.; Joo, H. G.; Jang, Y. H. Effects of Carbon Nanotubes on Electrical Contact Resistance of a Conductive Velcro System Under Low Frequency Vibration. *Tribol. Int.* **2016**, *104*, 45–56.

23. Armentano, I.; Dottori, M.; Fortunati, E.; Mattioli, S.; Kenny, J. Biodegradable Polymer Matrix Nanocomposites for Tissue Engineering: A Review. *Polym. Degrad. Stab.* **2010**, *95*, 2126–2146.

24. Spitalsky, Z.; Tasis, D.; Papagelis, K.; Galiotis, C. Carbon Nanotube–Polymer Composites: Chemistry, Processing, Mechanical and Electrical Properties. *Prog. Polym. Sci.* **2010**, *35*, 357–401.

25. Laredo, E.; Grimau, M.; Bello, A.; Wu, D.; Zhang, Y.; Lin, D. AC Conductivity of Selectively Located Carbon Nanotubes in Poly (ε-caprolactone)/Polylactide Blend Nanocomposites. *Biomacromolecules* **2010**, *11*, 1339–1347.

26. Kuan, C-F.; Kuan, H-C.; Ma, C-CM.; Chen, C-H. Mechanical and Electrical Properties of Multi-wall Carbon Nanotube/Poly (Lactic Acid) Composites. *J. Phys. Chem. Solids* **2008**, *69*, 1395–1398.

27. Sun, Y.; He, C. Synthesis, Stereocomplex Crystallization, Morphology and Mechanical Property of Poly (Lactide)–Carbon Nanotube Nanocomposites. *RSC Adv.* **2013**, *3*, 2219–2226.

28. Shao, S.; Zhou, S.; Li, L.; Li, J.; Luo, C.; Wang, J.; et al. Osteoblast Function on Electrically Conductive Electrospun PLA/MWCNTs Nanofibers. *Biomaterials* **2011**, *32*, 2821–2833.

29. Zhang, D.; Kandadai, M. A.; Cech, J.; Roth, S.; Curran, S. A. Poly (L-lactide) (PLLA)/Multiwalled Carbon Nanotube (MWCNT) Composite: Characterization and Biocompatibility Evaluation. *J. Phy. Chem. B* **2006**, *110*, 12910–12915.

30. Lee, H. H.; Sang, S. U.; Lee, J. H.; Kim, H. W. Biomedical Nanocomposites of Poly (Lactic Acid) and Calcium Phosphate Hybridized with Modified Carbon Nanotubes for Hard Tissue Implants. *J. Biomed. Mater. Res. Part B: Appl. Biomater.* **2011**, *98*, 246–254.

31. Ceregatti, T.; Pecharki, P.; Pachekoski, W. M.; Becker, D.; Dalmolin, C. Electrical and Thermal Properties of PLA/CNT Composite Films. *Revista Matéria* **2017,** 22, e11863.

32. Supronowicz, P.; Ajayan, P.; Ullmann, K.; Arulanandam, B.; Metzger, D.; Bizios, R. Novel Current-conducting Composite Substrates for Exposing Osteoblasts to Alternating Current Stimulation. *J. Biomed. Mater. Res. Part A.* **2002,** *59,* 499–506.

33. Yoon, O. J.; Kim, H. W.; Kim, D. J.; Lee, H. J.; Yun, J. Y.; Noh, Y. H.; et al. Nanocomposites of Electrospun Poly [(D, L-lactic)-co-(glycolic acid)] and Plasma-Functionalized Single-walled Carbon Nanotubes for Biomedical Applications. *Plasma Process. Polym.* **2009,** *6,* 101–109.

34. McCullen, S. D.; Stano, K. L.; Stevens, D. R.; Roberts, W. A.; Monteiro-Riviere, N. A.; Clarke, L. I.; et al. Development, Optimization, and Characterization of Electrospun Poly (Lactic Acid) Nanofibers Containing Multi-walled Carbon Nanotubes. *J. Appl. Polym. Sci.* **2007,** *105,* 1668–1678.

35. McCullen, S. D.; Stevens, D. R.; Roberts, W. A.; Clarke, L. I.; Bernacki, S. H.; Gorga, R. E.; et al. Characterization of Electrospun Nanocomposite Scaffolds and Biocompatibility with Adipose-derived Human Mesenchymal Stem Cells. *Int. J. Nanomed.* **2007,** *2,* 253.

36. Li, Z.; Zhao, X.; Ye, L.; Coates, P.; Caton-Rose, F.; Martyn, M. Structure and Blood Compatibility of Highly Oriented PLA/MWNTs Composites Produced by Solid Hot Drawing. *J. Biomater. Appl.* **2014,** *28,* 978–989.

37. Oliveira, J. E.; Mattoso, L. H. C.; Medeiros, E. S.; Zucolotto, V. Poly (Lactic Acid)/ Carbon Nanotube Fibers as Novel Platforms for Glucose Biosensors. *Biosensors* **2012,** *2,* 70–82.

38. Tsou, C-H.; Yao, W-H.; Lu, Y-C.; Tsou, C-Y.; Wu, C-S.; Chen, J.; et al. Antibacterial Property and Cytotoxicity of a Poly (Lactic Acid)/Nanosilver-Doped Multiwall Carbon Nanotube Nanocomposite. *Polymers* **2017,** *9,* 100.

CHAPTER 9

3D IMAGE PROCESSING IN STRUCTURAL CHARACTERIZATION OF ELECTROSPUN NANOFIBROUS MEMBRANES

BENTOLHODA HADAVI MOGHADAM[1], SHOHREH KASAEI[2], and A. K. HAGHI[1*]

[1]Department of Textile Engineering, University of Guilan, Rasht, Iran

[2]Department of Computer Engineering, Sharif University of Technology, Tehran, Iran

*Corresponding author. E-mail: akhaghi@gmail.com

ABSTRACT

Nanoporous membranes are an important class of nanomaterial that can be used in many applications, especially in micro- and nanofiltration. Electrospun nanofibrous membranes have gained increasing attention due to the high porosity, large surface area per mass ratio along with small pore sizes, flexibility, and fine fiber diameter, and their production and application in development of filter media. Image analysis is a direct and accurate technique that can be used for structural characterization of porous media. This technique, due to its convenience in detecting individual pores in a porous media, has some advantages for pore measurement. The three-dimensional (3D) reconstruction of porous media, from the information obtained from a two-dimensional (2D) analysis of photomicrographs, is a relatively new research area. This chapter provides a detailed review on relevant approach of 3D reconstruction from two views of single 2D image. The review concisely demonstrated that 3D reconstruction consists of three steps which is equivalent to the estimation of a specific geometry group. The method's algorithm has five major steps: (1) take a set of SEM images of a sample, (2) identify key points in the images that can possibly be detected in other images in the set, (3) search for corresponding points in images, (4) finding the fundamental matrix and computing 3D coordinates, and (5) triangulating the 3D points to compute 3D surface model. The obtained properties such as roughness, thickness, and porosity using proposed methods have been compared with the results of direct measurement. The evidence showed that the proposed method will be useful for extraction of volume information, such as porosity and thickness of nanofibrous structures with high precision and accuracy.

9.1   INTRODUCTION

Nanofibrous media have low basis weight, high permeability, and small pore size that make them appropriate for a wide range of filtration applications. In addition, nanofiber membrane offers unique properties such as high specific surface area (ranging from 1 to 35 m^2/g depending on the diameter of the fibers), good interconnectivity of pores, and potential to incorporate active chemistry or functionality on the nanoscale. During the past decades, several techniques have been proposed for characterization of morphology and properties of nanofibrous mats, such as diameter, roughness, thickness, porosity, contact angle, and alignment. The characterization of electrospun nanofibrous mat, due to their submicron size and random orientation in mat, are challenging.[1-6]

The roughness of nanofibrous mats has been characterized by several techniques such as *atomic force microscopy* (AFM), *angular-resolved scatter* (ARS), *total integrated scatter* (TIS), X-ray reflectivity, profilometer (contact and noncontact mode), *scanning electron microscopy* (SEM), entropy method, and image processing techniques. Among these methods, the contact profilometer is very simple and sufficient technique for roughness measurement. However, it suffers from the physical contact of the stylus tip with the sample surface, which disrupts the sample surface and causes an error in the characterization. Measuring the roughness of nanofibrous structures by AFM has been suggested by some researchers.[7-15] Baltsavias compared two different approaches of roughness measurement. The results show that the indirect method is more practical and useful for calculation of standard surface profile measurement, compared with the stylus profilometer for characterization of a surface topography.[14] Nowadays, measuring roughness of different surfaces by AFM has gained much interest, due to the fact that AFM is essentially an unsurpassed surface measurement instrument. The AFM method has many capabilities, such as the characterization of surface and permitting 3D measurement at higher lateral and vertical resolution, than profilometer method.[13–19]

The thickness of nanofibrous mat is measured by some apparatus such as the micrometer, profilometer, microscope (cutting of nanofibrous mat and viewing the cross-section), contactless laser measurement, and ultrasonic technology. However, these techniques suffer from some weaknesses such as systematic error and require expensive devices. Measuring the thickness of nanofibrous mat, due to their soft and multilayer structure with unknown boundaries of the layers and nonuniform thicknesses in different parts of the mat, is a big challenge for researchers. Fortunately, the application of image analysis techniques can be helpful in finding and optimizing the electrospun nanofibrous mat characteristics. According to the related literature, the two-dimensional (2D) image analysis of scanning electron microscope micrographs for geometrical characterization has been used for measuring the total porosity, pore size distribution, and diameter of relatively thin nonwovens. However, due to its limitations on relatively small field-of-view and its restricted information about the whole structure, it cannot be used for characterization of multilayer electrospun fibrous mat.[20–23]

Characterization techniques for evaluating the performance of any porous structure can be classified into microscopic and macroscopic techniques. Microscopic techniques usually consist of high-resolution light microscopy, SEM, and micro-computed tomography (Micro-CT). The main problem of these time-consuming and expensive techniques is that they cannot determine flow properties. Various techniques may be used for assessment of the pore characteristics of porous membranes through macroscopic approach, including mercury intrusion porosimetry, liquid extrusion flow porometry, liquid extrusion porosimetry, and pycnometry. The low stiffness and high pressure sensitivity of nanofibrous mats limit the application of these techniques. Hence, an accurate estimation of porosity of nanofibrous mat is difficult, so that, sometimes a combination of techniques is required.[24–33]

To measure the pore characteristics of electrospun nanofibrous membranes using image analysis, images (or micrograph) of the nanofiber

webs, which are usually obtained by SEM, transmission electron microscopy (TEM) or AFM, are required. Image analysis is a direct and accurate technique that can be used for characterization of porous media. This
technique, due to its convenience in detecting individual pores in a porous
media, has some advantages for pore measurement.[34] In the meantime, the
application of image analysis techniques can be helpful in an effort to find
and optimize the electrospun nanofibrous mat characteristics. According
to the literature, 2D image analysis of *scanning electron microscope*
micrographs for geometrical characterization has been used for measuring
the total porosity, pore size distribution, and diameter of relatively thin
nonwovens. However, due to its limitation on relatively small fields of
view as well as its restricted information about the whole structure, it
cannot be used for characterization of multilayer electrospun fibrous mat.
Nanofibrous membranes typically contain two kinds of structure, 2D or
three-dimensional (3D) nanofibrous network assemblies while the 3D
structure provides good handling characteristics (Fig. 9.1). Hence, in this
work, the 3D reconstruction of nanostructure is considered to significantly
improve the prediction of the depth and dimension of structures.

There are several instrumental characterization techniques to obtain
3D images of nanofibrous mat, such as Micro-CT,[35] nondestructive 3D
laser scanning confocal microscope (LSCM),[36] 3D electron back-scatter
diffraction (EBSD),[37–39] and focused ion beam-scanning electron microscope (FIB-SEM).[40,41] However, these techniques are limited by their
resolution. Thus, the 3D stochastic reconstruction of porous media from

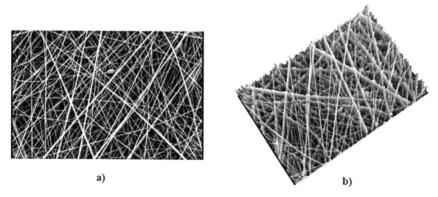

a)   b)

FIGURE 9.1   SEM image of electrospun nanofibrous membrane: (a) 2D and (b) 3D.

statistical information of 2D photomicrographs has been suggested by some researchers.[42]

Recently, researchers have focused on the 3D image analysis for obtaining 3D models from images. The 3D reconstruction of porous media, from the information obtained from a 2D analysis of photomicrographs, is a relatively new research area. Due to the complexity of images, conventional modeling techniques are very time consuming and thus recreating the detailed geometry becomes very difficult. In order to overcome these difficulties, some researches have inclined toward image-based modeling techniques to drive the 3D reconstruction.[43–50] Over the years, there have been few studies directed toward the 3D reconstruction of nanofibrous membranes. Kazemi et al.[51] defined a model for the nanostructured fibrous network as an ideal membrane using adaptive local criteria in the image analysis. In this study, the images are captured from different regions of a membrane and the number of pixels within a distinct grayscale level is calculated by selecting a specific grayscale interval. Each local subimages are determined using the threshold values based on neighboring pixels within a specific radius. Sambaer et al.[52] proposed the realistic 3D structure model, as mutually connected tubes having different diameters, of produced polyurethane (PU)-based nanofiber electrospun mat. The 3D structure obtained using rotation in the depth of particular average diameter circle along every corresponding centerline pixel. Jaganathan et al.[53] obtained the 3D geometries based on fiber-level information via digital volumetric imaging (DVI) technique. Also, the 3D geometries are converted into stereo lithography (STL) files for meshing such geometries using a computer-aided design (CAD) technique. Zobel et al.[54] in order to 3D reconstruction of fiber-webs assumed that fibers have square cross-sections and bend over each other according to a simple set of rules. Faessel et al.[55] developed a 3D model of random fiber networks using the Visualization Toolkit (VTK) libraries. In order to obtain a finite element mesh of a unit volume of the material, the 3D model of the network is discretized by shell elements. Also, morphological data of real networks extract from X-ray micro-tomographic observations. Soltani et al.[56] proposed the realistic 3D images of fibrous networks using sets with various degrees of alignment including nearly isotropic, nearly layered, and moderately aligned networks. Also, the direct 3D model of the needled nonwoven fabrics generated by X-ray Micro-CT is used in conjunction with a computational fluid dynamics model for the simulation of transverse permeability. Hosseini et al.[57–58] generated 3D fibrous geometries resembling the microstructure of a fibrous medium, by C++ computer program to produce fibrous structures of different fiber diameters, porosities, thicknesses, and orientations. Ji et al.[59] used for 3D reconstruction of the nanofibrous scaffold from LSCM. Reingruber et al.[60] extracted the 3D information about the membrane structure from serial sectioning and imaging of the membrane embedded in all three directions. Ostadi et al.[61] generated a binary 3D model of the complex structures using software techniques and threshold tuning the grayscale X-ray images. The actual 3D structure of the fuel cell is captured through X-ray and FIB/SEM nanotomography. The abovementioned methods are very expensive, time consuming, and do not have readily accessible instruments. Also, recreating detailed geometry becomes very difficult which cannot be fully automated, which itself poses a substantial source of errors. However, among these methods, 3D reconstruction from statistical information is the most attractive option for characterization of nanofibrous membranes. 3D reconstruction of a number of perspective images is a challenging task in computer vision due to the loss of depth in the process of the photographing image. The 3D reconstruction of 2D images, from the information obtained from a 2D image analysis, is a relatively new research area.[62–69]

Recently, the problem of interactive 3D reconstruction of a number of perspective images is one of the fundamental problems of computer vision while reconstruction from two views from single image is the simplest one. The 3D reconstruction steps from two views for a 2D image are shown in the following stages (Fig. 9.2)[70]:

(a) Detection of feature points in two images;

(b) Finding matched points in two views of images;

(c) Triangulating the 3D points into a 3D mesh;

(d) Computing 3D points from 2D matched points; and

(e) Mapping a 2D image as a texture on the surface.

Finding the 2D matched points between two images is the first challenge in the process, whereas image matching is a fundamental problem in computer vision. Matching can provide valuable information about the similarity between the images for 3D reconstruction from multiple

Input images

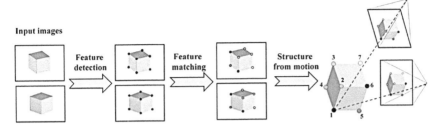

FIGURE 9.2   3D camera poses and positions.

images. The second challenge is the determination of the 3D location of each image point. If we know the calibration parameters of the camera but not the point positions, we could find the points through triangulation; conversely, if we do not know the calibration parameters of the camera, we cannot write the projection relations using normalized image but we have to include the calibration matrixes.[71–75]

The method's algorithm for a 2D image pair is based on the following stages (Fig. 9.3)

This chapter is organized as follows. In the Introduction section, we introduce the nanofibrous membranes and structural characterization approaches such as the roughness, thickness, porosity measurement, and summarize recent literature review. In this section, we have proposed the best and effective method of structural characterization of nanofibrous membranes. In Section 9.2, a novel technique to assess the roughness of nanofibrous mats based on image processing inspired by the simple assumption based on its grayscale variations is proposed. The height of nanofibrous mat in different regions of the surface is simulated by grayscale variations in the image while the relation between grayscale and height is obtained as a linear function. The roughness is obtained by measuring height variations in the surface profile. Also, we have proposed a novel approach based on 3D reconstruction from two views of single 2D SEM image for thickness and porosity measurement of electrospun nanofibrous membranes. The proposed method exhibits very realistic 3D surfaces for 3D visualization of nanofibrous membranes. The performance of the proposed method has been demonstrated on different regions of nanofibrous membranes. The resulting points were reconstructed and the errors were evaluated and discussed. Finally, the high accuracy was

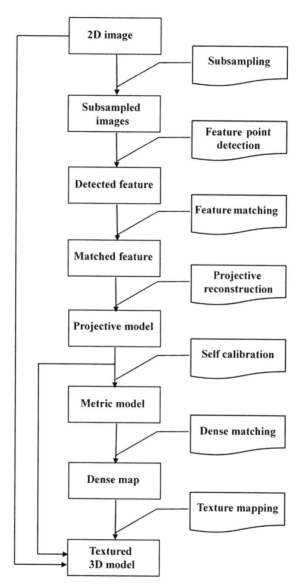

FIGURE 9.3   3D modeling algorithm.

obtained in the low magnification and small angle between two viewpoints of SEM image. The thickness of three nanofibrous mats, including polyacrylonitrile (PAN), polyvinyl alcohol (PVA), discussed in this chapter , and PU are calculated using different views (15°, 30°, and 45°) of 2D images captured by a fixed scanning electron microscope and rotating samples in several positions. By calculating the disparity value (the distance between two corresponding points in two images), the relative depth of image and consequently, the thickness of nanofibrous mat are obtained. For calculation of porosity by using the results of the previous section, the ratio of the pore volume to the total volume of nanofibrous mat is defined as the ratio of the number of black voxels to the total number of voxels in the 3D image. An intermediate Section 9.3 establishes the verification step for any kinds of properties. Statistical roughness parameters of nanofibrous mats are also obtained from direct measurement with roughness profilometry and AFM for comparison of the roughness of nanofibrous mats between direct measurement and proposed method. The AFM is the most suitable surface measuring instrument for roughness measurement on the nanofibrous mat. Also, the thickness of three electrospun mats is measured from the cross-sectional view of nanofibrous mat by SEM. The obtained porosity has been compared with the results of water pycnometry and 2D image processing method. Finally, Section 9.4 concludes the chapter.

9.2   APPLICATION OF IMAGE ANALYSIS IN STRUCTURAL CHARACTERIZATION OF NANOFIBROUS MEMBRANES

9.2.1   ROUGHNESS MEASUREMENT

SEM images are usually used as an input for characterization of nanofibrous mats via image processing techniques. A 3D surface roughness evaluation has been established based on the digital image processing technology using surface profile information.[76–80] In most of the cases, image processing-based techniques formulate the roughness using probability distributions of random variables. Deterministic methods offer an alternative device to measure the roughness by using nonstatistical approaches. The comparison of various image processing methods is summarized in Table 9.1.

TABLE 9.1 Comparison of Image Processing-based Techniques of Roughness Measurement.

Measuring method	Description	Advantages	Disadvantage	Error and accuracy	References
Image processing based on grayscale calculation	Determine the correlation between some parameters by artificial neural network	Fast, reliable without considering the setting of the camera	Variation of the results reached 33% in several cases	Reliability and accuracy resulted data depended to a great extent on such parameters	76
Digital image processing	A three-dimensional surface roughness evaluation system consisting of both hardware and software architecture	The vision is more perfect. The image quality had an obvious improvement	-	Detail is more accurate	77
O-IDIP method	Quantify surface roughness on a comparative basis using Std Dev L measurements	Application at various scales, from sensor to the building scale	It require more adjustments of contrast	-	78
Vision system-based image processing technique	The Euclidean and Hamming distances of the surface images are used for surface recognition	Online and noncontact method	-	When a large database of reference images covering the entire range of surface roughness values is maximum accuracy	79
Dividing of the histogram peak by line between the darker and lighter pixels	The standard deviation of the pixel brightness was called the "SEM roughness index" and calculated for each surface	Able to detect small changes in x–y uniformity	The technique is unlikely to find use as a real-time quality control tool	-	80

O-IDIP, digital photography and image processing.

It is worth noting that the roughness measurement using image processing methods has many advantages, including fast measurement, noncontact measurement, inexpensiveness, high information content, and surface measurement capability. Hence, this chapter presents a novelty method as simple function based on image processing for determination of height variations inspired by the simple assumption that the grayscale values equal to 0 and 255 are considered as the minimum and maximum height of the nanofibrous mats. This aims to obtain a relation between grayscale variations in the image and the real sample height variations as a linear function. Finally, the obtained results are compared with the direct measurement of nanofibrous mat roughness through profilometer and AFM results.

9.2.1.1 PROPOSED METHOD

Image-processing method was used in this work for finding simple and effective function for assessment of height in different part of nanofibrous mat surface, so that roughness parameter can be extracted from height profile of the surface. Figure 9.4 shows the flowchart of the proposed method.

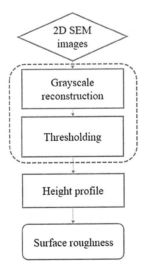

FIGURE 9.4 Flowchart of proposed method.

A scanning electron micrograph is used for image acquisition. Grayscale reconstruction with binary SEM image can be used to identify local extrema in the image and followed by the assessment of surface roughness in the image.[81-85] The peaks and valleys heights and the peak to valley ratio can be calculated from the calibrated grayscale histogram. The grayscale histogram contains many important information such as the minimum grayscale (I_{min}), maximum grayscale (I_{max}), average grayscale numbers (μ), and the most frequent intensity (P). The average grayscale numbers (μ) is calculated by

$$\mu = \frac{\sum_{i=0}^{255} np(n)}{\sum_{i=0}^{255} p(n)} \tag{9.1}$$

where n is the grayscale number and $p(n)$ represents number of counts for the nth grayscale number.[10,86-90]

For obtaining the structural details about 2D SEM images, it is necessary to define the correct threshold level (χ), which can be determined from the grayscale histogram, according to

$$\chi = \frac{(I_{max} - I_{min}) \times \mu}{P} \tag{9.2}$$

Then, the given threshold is applied to all SEM images to perform the height distribution analysis.

In this study, following assumptions are made:

Two calibrated points are used as an input values.

Grayscale values equal to 0 and 255 are considered as the minimum and maximum height of the nanofibrous mats, respectively (*see* Fig. 9.5).

The profilometer device is used to measure the surface roughness. According to this method, a good general description of height variations is provided to surface roughness measurement (Fig. 9.6).[19]

It can be shown that the average surface roughness value (Ra) is closely related to the vertical distance between the lowest and highest height of the profilometer.[91-94] The root mean square roughness (Rq) is calculated from the standard deviation of the height data according to eq 9.4 to describe the deviation of the measurement points to the mean.

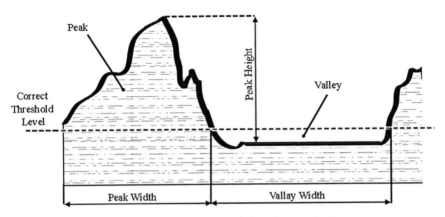

FIGURE 9.5   Definition of structural characteristics of grayscale histogram.

$$R_a = \frac{\sum_{i=0}^{n} |Z_i - Z_m|}{n} \qquad (9.3)$$

$$R_q = \frac{\sqrt{\sum_{i=0}^{n} (Z_i - Z_m)^2}}{n} \qquad (9.4)$$

$$z_i = \frac{z_{max} - z_{min}}{I_{max}} \times I_i + z_{min} \qquad (9.5)$$

where n = evaluation length, Z_i = height, Z_m = mean height.

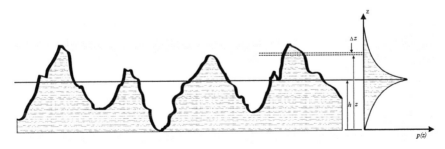

FIGURE 9.6   Profile height distributions.

9.2.1.2 EXPERIMENTAL RESULTS

9.2.1.2.1 Materials and Methods

PAN (MW = 100,000 g/mol, Polyacryle Co., Iran), and *N-N*, dimethylformamide (DMF, Merck Co., Germany) were used.

The PAN powder was added to DMF and stirred by a magnetic stirrer for 24 h at 25°C to obtain a homogeneous solution. The polymer solution was an electrospun using an electrospinning apparatus consisting of a glass syringe, a syringe pump, a ground electrode, and a high-voltage power supply. To produce electrospun nanofibrous mats, the polymer solution with two different concentrations (i.e., 10 and 12 wt.%) was put into a plastic syringe with a steel needle tip as a nozzle for electrospinning. The polymer solution was fed to the needle tip at the injection rate of 2 mL/h by a syringe pump. The high voltage (22 kV) was connected to the external surface of the metallic needle and the tip to fixed grounded collector distance was 10 cm.

9.2.1.2.2 Characterization Techniques

The SEM (Philips XL–30) was used for morphology evaluation of the nanofibrous mats. Samples were coated with a thin layer of gold using a sputter coater. Due to the random structure of nanofibrous mats, the SEM image of the sample was obtained from several regions of nanofibrous mats. The *average fiber diameter* (AFD) and the diameter distribution were determined from the SEM image by the used measurement software.

The roughness of the nanofibrous mats was also characterized by a contact or stylus profilometer device (Veeco Dektak Series 3, USA) with vertical range of 10 nm–100 μm. This device is capable of measuring the heights of samples with nanometer resolution in a contact manner. Furthermore, the topography and surface profile of the nanofibrous mat were analyzed using AFM (Ara-A.F.M. Model No.0101/A, Iran). All experiments were conducted in air and room temperature and the CSC17 silicon probe from Mikromasch Company was employed.

9.2.1.3   RESULTS AND DISCUSSION

Determination of fiber diameters and its distribution were carried out manually. The experimental results revealed that the nanofibers diameter is increased by increasing the concentration of polymer solution (Fig. 9.7). In the higher solution concentration, there are more polymer chain entanglements and less chain mobility. Therefore, the jet extension would be disrupted during the electrospinning process and thicker fibers would be produced.[95]

The SEM image was provided for two nanofibrous mat samples with different diameters (187 and 213 nm). The binary image of the electro-spun nanofibrous mat was produced from the SEM images using local thresholding. As shown in Figure 9.8, the local thresholding algorithm was applied for SEM images which provide a binary image. The useful

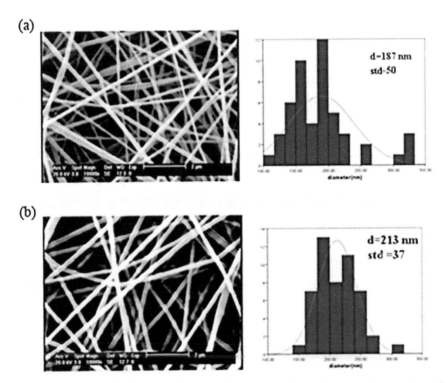

FIGURE 9.7 Morphology and diameters distribution of nanofibers at concentration of (a) 10 wt.% and (b) 12 wt.%.

information about the SEM image was provided by the graylevel histogram. The threshold is calculated using eq 9.2 in various parts of the image. In order to find an appropriate threshold level, it is necessary to extract the given parameters from these histograms (Table 9.2).[96–97]

TABLE 9.2 Extracted Parameters from Grayscale Histogram.

Sr. no.	Diameter (nm)	I_{min}	I_{max}	μ	P	χ
1	187	0	255	65.07 62	53,590	0.31
2	213	0	255	55,78.93	79,608	0.25

Line profile of grayscale shows the variations of grayscale along a line. Line profiles are helpful for finding height variations in different parts of the image. The change in brightness along the horizontal line indicates a height variation in SEM images. The profile graphs corresponding to the

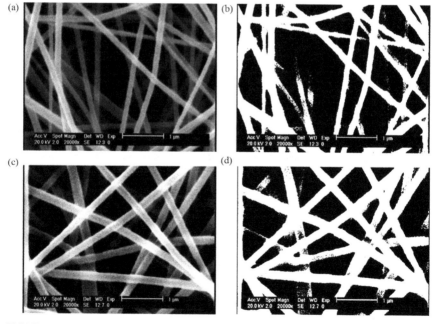

FIGURE 9.8   SEM (a and c) and binary (b and d) images of nanofibrous mat with $d = 187$ nm (a and b), and $d = 213$ nm (c and d).

grayscale variation are shown in Figure 9.9. The images were normalized to have uniformity in image pixel grayscale.[97]

It is necessary to use grayscale characteristics for estimation of the height of nanofibrous mats in several part of surface. Consequently, the height of sample (Zmin, Zmax) was measured by profilometer and then two given points were calibrated (grayscale value is equal to Zmin = 0 and Zmax = 255). By using these two calibration points, the whole range of the grayscale values of the SEM images (*I*) has been related to the measured height of sample (*z*), considering a linear relation between these two variables.

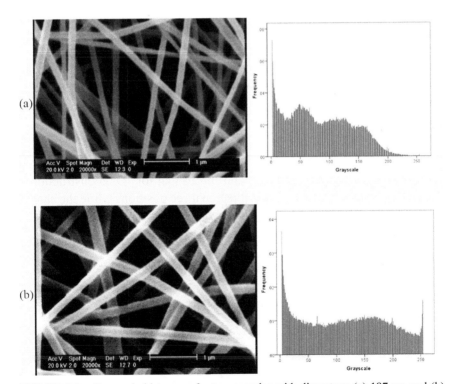

FIGURE 9.9   Grayscale histogram for two samples with diameters: (a) 187 nm and (b) 213 nm.

$$z_i = \frac{z_{max} - z_{min}}{I_{max}} \times I_i + z_{min} \qquad (9.5)$$

9.2.1.4   DIGITAL SURF MOUNTAINMAP SOFTWARE

Once the 3D dataset is acquired from each optical instrument, it is then processed using the Digital Surf MountainMap (v5) software (instead of using each instrument's native software). Instrument native software processing of 3D data is often subject to issues of repeatability and may be nontraceable in nature, leading to significant uncertainty about data quantitation and representation. Moreover, for better intercomparison of these instruments, using the third party software for data analysis is an appropriate approach.

Hence, due to the sensitivity of the issue, in order to overcome these problems, we have provided a novel, precise, and economical technique based on the 3D model from two views of the single SEM image for determination of height variation inspired by the analytical software "Mountains Map®SEM" in different parts of nanofibrous mat surface, so that roughness parameter can be extracted from height profile of the surface.

9.2.1.4.1 Proposed Method

In this work, we have prepared two different views in three different directions in SEM images of the nanofibrous mat for 3D reconstruction of single 2D images by rotating the object in three positions while the camera is fixed (Fig. 9.10).

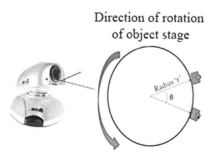

FIGURE 9.10   Proposed setup to capture images.

Geometrical structure of objects is in location of corresponding points that allows calculating the height according to eq 9.6, in which α is the angle between the two views. This is also the geometry that applies when a single viewpoint is used and the object is rotated to obtain two images. The scanning electron microscope typically uses that procedure to capture images,[98] as

$$Z = \frac{(d_1 - d_2)}{2 \sin \frac{\alpha}{2}} \qquad (9.6)$$

where d_1, d_2 are the disparities between pixels and Z is a relative height in different regions of sample.

The surface profile is used to measure the surface roughness. According to Figure 9.11, the height variations are provided from surface profile to surface roughness measurement.

Also, the average surface roughness value (Ra) is closely related to the vertical distance between the lowest and highest height of the 3D profile. The root mean square roughness (Rq) was calculated from the standard deviation of the height data according to eqs 9.3 and 9.4 to describe the deviation of the measurement points to the centerline.[78,91–93]

9.2.1.4.2 Results and Discussion

At the beginning, different views of the PMMA/PAN nanofibrous mats were prepared from single 2D SEM images (Fig. 9.12).

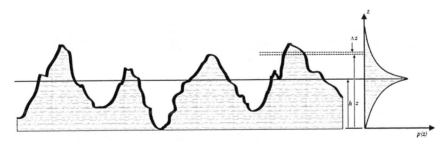

FIGURE 9.11 Profile of height distribution.

Rotated angle(°)	Original image	Rotated image
15		
30		
45		

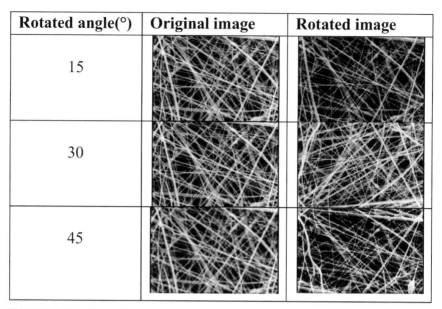

FIGURE 9.12   Three different views of nanofibrous mat.

Then, data processing using the MountainMap software to manually evaluate the height of indentations is used in this research. Typically, the 3D datasets also contained the geometric form that is needed to be removed for better assessment of the surface. The surface topography can also be obtained by using this technique (Fig. 9.13).

The surface topography is reproduced as a 2D profile and it is truly representative of the surface. The surface profile gives information on the morphology of the surface texture so that the positive values show high peaks spread on a regular surface and negative values show surfaces with pores and scratches (Fig. 9.14).

The average roughness of the nanofibrous mat can be rapidly determined by eqs 9.3 and 9.4 as shown in Table 9.3.

TABLE 9.3 Average Roughness of the Nanofibrous Mat in Three Different Views.

Rotated angle (°)	Average surface roughness (µm)
15	0.13
30	0.45
45	0.75
Mean	0.44

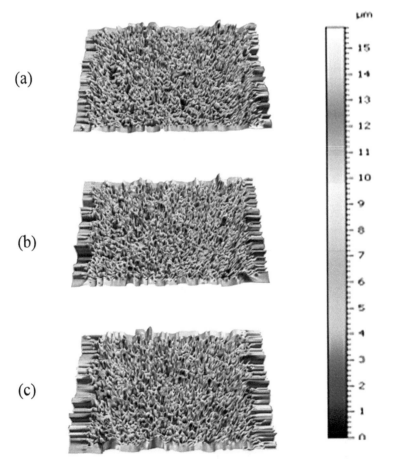

FIGURE 9.13 **(See color insert.)** Surface topography in three different views: (a) 15°, (b) 30°, and (c) 45°.

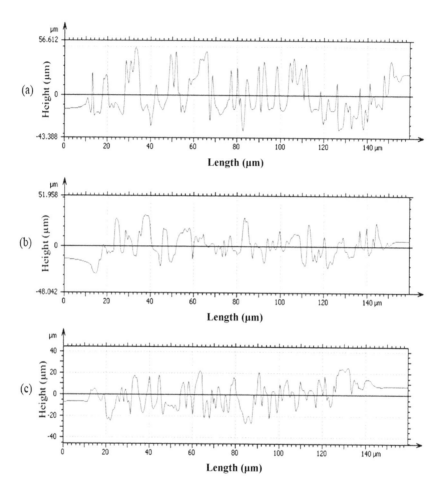

FIGURE 9.14 Profile of height distribution in three different views: (a) 15°, (b) 30°, and (c) 45°.

9.2.2 THICKNESS MEASUREMENT

Depth estimation is a fundamental subject in 3D image analysis and has a critical role in 3D reconstruction. It provides detailed information about the thickness of structures. There are various depth estimation techniques from single 2D image, such as linear perspective, texture gradient, overlapping, and relative height. Among the vision-based techniques, most work has been focused on stereovision and other algorithms based on multiple images, such as optical flow, structure-from-motion, and depth from defocus. There are also some algorithms for depth estimating from single image in very specific settings, such as shape-from-texture and shape-from-shading.[99–105] For instance, Sambaer et al.[42] used an image analysis technique for thickness measurement with a micrometer on the real nonwoven sample. They found the relation between the grayscale level of images and the real sample thickness. In another study, the thickness of nanofibrous mat was calculated after 3D reconstruction of images by using the confocal laser scanning microscopy.[37] But, to the best of our knowledge, there are no reports on utilization of depth information for predicting the thickness of nanofibrous mat. Hence, in this work, a study was conducted to investigate the depth of SEM images of three nanofibrous mats of PVA, PAN, and PU, by using the information of two views of the 2D SEM image. First, two views of the image are captured by rotating the sample in three angles (i.e., 15°, 30°, and 45°) and then the depths of images are obtained through the disparity calculation. Finally, the obtained results are compared with direct measurement of nanofibrous mat thickness through cross-sectional imaging to show the efficiency of the proposed method.

9.2.2.1   PROPOSED METHOD

Figure 9.15 shows the flowchart of the proposed method for thickness estimation. The processing is carried out in several subsequent steps. First, the corresponding points with known coordinates on the two views of each nanofibrous mat are detected. Then, the disparity (the difference of distances between different coordinates of the points) is calculated. This is then used to generate the 3D point coordinate values and depth of images.[106–109]

9.2.2.1.1   Stereo Matching

The corresponding points in two image views are defined by using stereo matching. The *scale-invariant feature transform* (SIFT) key point extractor is used to find the corresponding points. The scale-invariant features are identified by using four major processing steps: (1) scale-space peak selection; (2) key point localization; (3) orientation assignment; and (4) key

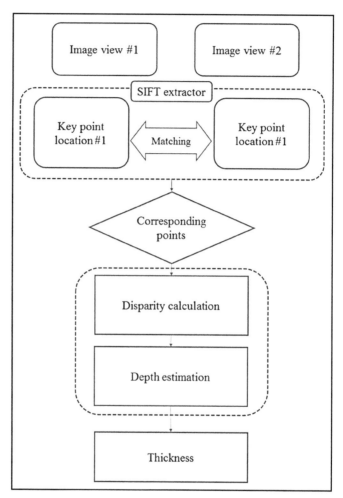

FIGURE 9.15   Flowchart of proposed thickness estimation method.

point description. The magnitude and orientation of image gradients are sampled around the key point location using the region scale to select the level of Gaussian blurs. This descriptor focuses on the issue of robustness to lighting variations and small positional shifts by encoding the image information in a localized set of gradient orientation histograms. The large numbers of extracted features in the full range of scales and locations is an important advantage of this method.[110–112]

9.2.2.1.2 Disparity Calculation

The relative depth of samples can be obtained by calculating the disparity between the corresponding pixels presented at different views (Fig. 9.16). The disparity is defined as a displacement of the matched pixels in two different views.[113–115]

9.2.2.1.3 Depth Estimation

The depth information can be obtained from the corresponding disparity measures by using microscope information.[99–105] The depth is inversely proportional to disparity, defined by

$$Z = \frac{fB}{d} \tag{9.7}$$

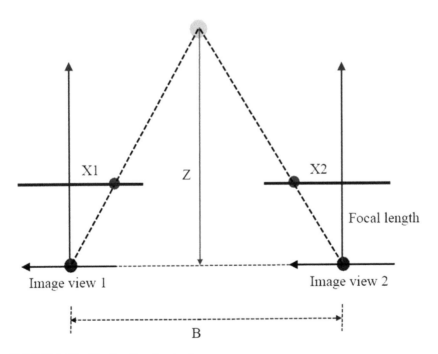

FIGURE 9.16   Depth estimation basics.

$$B = \frac{1}{\cos(\theta)} \tag{9.8}$$

$$d = X_2 - X_1 \tag{9.9}$$

where B indicates the distance between two positions of a sample, θ is the view angle between two images, d is the disparity between corresponding pixels, Z indicates the depth (distance of the sample from the microscope), and X_1, X_2 are corresponding image points. From eq 9.7, it can be concluded that a sample with lower depth exhibits a higher disparity between the corresponding pixels in an image. In other words, disparity is high for a sample close to the microscope.[113–115]

9.2.2.1.4   Thickness Calculation

The thickness of nanofibrous mat is obtained by converting the depth pixel to the thickness value. The differences in thickness of each point can be attributed to the random orientation of the nanofibrous mat and the number of nanofibers layers.

9.2.2.2   EXPERIMENTAL RESULTS AND DISCUSSIONS

9.2.2.2.1   Materials and Methods

The PAN powder with weighted average *molecular weight* (Mw) of 100,000 g/mol was obtained from Polyacryle Co. (Iran). PVA (Mw = 72,000 g/mol), PU (Mw = 18,000 g/mol), DMF, and tetrahydrofuran were provided by Merck Co. (Germany). These chemicals are commonly used without further purification.

The polymers were added to the solvent and stirred for 24–30 h at room temperature to obtain a homogeneous solution. Then, the electro-spun nanofibrous mats were prepared according to the electrospinning condition tabulated in Table 9.4. The electrospinning apparatus used in this experiment was produced by Fanavaran Nano Meghyas Co. (Iran).

TABLE 9.4 Electrospinning Conditions.

Sample	Solvent	Concentration (wt.%)	Voltage (kV)	Spinning distance (cm)	Injection rate (mL/h)
PAN	DMF	10	14	10	2
PVA	Water	15	18	18.5	0.5
PU	DMF/THE (1.5/1)	12	15	15	0.5

DMF, N-N, dimethylformamide; PAN, polyacrylonitrile; PU, polyurethane; PVA, polyvinyl alcohol; THE, tetrahydrofuran.

9.2.2.2.2 Measurement

The micrographs of nanofibrous mat were obtained by using the SEM (Philips XL–30). First, the electrospun mat was coated with a thin layer of gold using a sputter coater. Several areas were captured in order to examine the uniformity of the mat thickness. Different views of the electrospun nanofibrous mat were obtained for thickness estimation from single 2D SEM images obtained by rotating the sample and the fixed scanning electron microscope.

9.2.2.3 OBTAINED RESULTS

The SIFT descriptor was applied on three different views (θ) of nanofibrous mat (see Fig. 9.17–9.19). As shown in Table 9.5, the number of matched points increases by decreasing the angle between two views. This is attributed to the fact that larger the angle between two views, the bigger the perspective distortions. As such, it is getting more difficult for the matching algorithm to correctly identify the corresponding points. These differences attribute the source of error in depth estimation (which tries to find the matches based on the assumption that corresponding points or edges must be identical (or very nearly so). Therefore, the angle of rotation should be kept as small as possible (15°) in order to overcome the problem of quick disappearance of objects in successive views.

TABLE 9.5   Number of Matched Points in Three Different Views of Samples.

Angle of rotation (°)	Number of regions	Number of matched points		
		PAN	PVA	PU
	1	357	250	654
	2	342	244	697
15	3	248	295	670
	4	335	407	670
	5	440	383	616
	1	68	138	209
	2	100	128	229
30	3	63	125	250
	4	88	178	161
	5	117	171	175
	1	54	119	90
	2	72	106	114
45	3	58	94	97
	4	64	109	84
	5	58	118	74

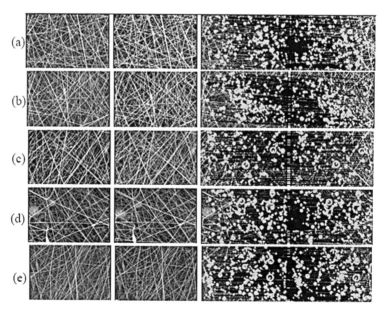

FIGURE 9.17   SIFT descriptor applied on two views of image in $\theta = 15°$ in five different regions of PAN nanofibrous mat.

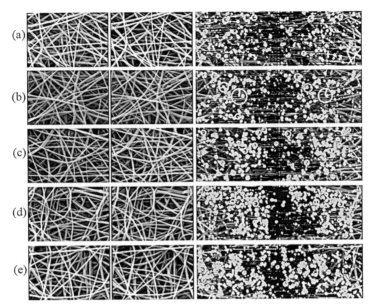

FIGURE 9.18   SIFT descriptor applied on two views of image in $\theta = 15°$ in five different regions of PVA nanofibrous mat.

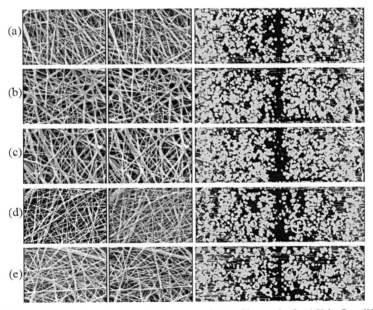

FIGURE 9.19   SIFT descriptor applied on two views of image in $\theta = 15°$ in five different regions of PU nanofibrous mat.

The proposed algorithm was applied on five different regions of nanofibrous mat by matching many points in the images in which the disparity between the points was calculated for all samples. It helps to compute the mean depth from disparity $(d_{min}$ and $d_{max})$ in two image views (Table 9.6).[100] The estimated depth is calculated as

$$Z_{mean} = \frac{Z_{max} + Z_{min}}{2} \tag{9.10}$$

By converting the pixel to the micrometer, the thickness of nanofibrous mat is obtained. Table 9.7 summarized the thickness of nanofibrous mat.

TABLE 9.6 Depth Estimation Parameters in Five Different Regions of Nanofibrous Mat at 15°.

Sample	Number of regions	Z_{max} (pixel)	Z_{min} (pixel)	Z_{mean} (pixel)
PAN	1	0.059	0.055	0.057
	2	0.070	0.066	0.068
	3	0.075	0.099	0.087
	4	0.050	0.064	0.057
	5	0.069	0.067	0.068
PVA	1	0.073	0.055	0.064
	2	0.074	0.054	0.064
	3	0.058	0.056	0.057
	4	0.055	0.081	0.068
	5	0.055	0.059	0.057
PU	1	0.093	0.073	0.083
	2	0.071	0.065	0.068
	3	0.061	0.067	0.064
	4	0.117	0.057	0.087
	5	0.051	0.053	0.052

PAN, polyacrylonitrile; PU, polyurethane; PVA, polyvinyl alcohol.

TABLE 9.7 Thickness of Nanofibrous Mat in Five Different Regions of Samples Obtained by Proposed Method.

Sample	Thickness of nanofibrous mat at different regions (µm)					Mean thickness (µm)
	1	2	3	4	5	
PAN	15	18	23	15	18	17.8 ± 3.27
PVA	17	17	15	18	15	16.34 ± 1.34
PU	22	18	17	23	14	18.8 ± 3.7

PAN, polyacrylonitrile; PU, polyurethane; PVA, polyvinyl alcohol.

9.2.3 PORE STRUCTURE ANALYSIS

Many characterization techniques have been applied to obtain 3D volume images of pore space, such as X-ray Micro-CT and magnetic resonance Micro-CT. However, these techniques may be limited by their resolution. So, the 3D stochastic reconstruction of porous media from statistical information (produced by analysis of 2D photomicrographs) has been suggested. 2D image analysis was unable to predict morphological characteristics of porous membrane, because it has insufficient information about the microstructure. Therefore, 3D reconstruction of porous structure will lead to significant improvement in predicting the pore characteristics. Recently, research work has focused on the 3D image analysis of porous membranes.[116–121]

For pore structure analysis applied the threshold to segment of binary image are pores (black voxels) and solids (white voxels). Voxels with a gray value higher than threshold value become white and voxels with a gray value smaller than a threshold will be black. A closed pore in 3D is a connected assemblage of space (black) voxels that is fully surrounded on all sides in 3D by solid (white) voxels. The threshold is given by the condition that the fraction of voxels with gray values that are smaller than the threshold must correspond to the porosity. For calculation of porosity by using the results of the previous section, the ratio of the pore volume to the total volume of nanofibrous mat is defined as the ratio of the number of black voxels to the total number of voxels in the 3D image.[122–123]

In the multiview approach, an image-based algorithm is employed for creating a 3D model of 2D images from two or more views. This method utilizes stereo pairs taken by tilting the sample stage with considering different perspectives from different view angles to restore the 3D structure of a specific object. These method has five major steps: (1) take a set of SEM images of an nanofibrous mat, (2) identify key points in the images that can possibly be detected in other images in the set, (3) match the images and computing the 3D matched points from the two stereo images, (4) use the projection geometry theory to estimate projection matrices, and (5) apply linear triangulation to compute 3D surface model and map a texture on the resulting 3D mesh.

The first step in our method is the extraction of key points (feature descriptors) and the second step is corresponding the extracted key points. To this purpose, pairs of key points are matched across three different views of the nanofibrous membrane. Thus, we must provide a way to find the best match in the other image. Finding the 2D matched points between two images is the first challenge in the process, whereas image matching is a fundamental problem in computer vision. Matching can provide valuable information about the similarity between the images for the 3D reconstruction process from multiple images.[124] The SIFT key point extractor is used to find the extract and describe the key points. An important advantage of this method is that it extracts a large number of key points to densely cover the image over the full range of scales. Note that, one needs at least seven corresponding points for reconstruction or eight points, if the problem is to be solvable linearly.[125]

The second challenge is the determination of the 3D location of each image point. For obtaining a 3D point cloud, it is necessary to compute the 3D position associated with each match. For this end, the Delaunay triangulation is an excellent way that is computed using the Computational Geometry Algorithms Library (CGAL). It provides our data points as grouped into sets of points that belong to the same global surface and evaluate the projection of each triangle in the chosen views. It also computes the mean of the color variance of the pixels in that triangle and obtains the 3D position of the key points to modify a generic model by using a geometrical deformation.[126,127]

If the calibration parameters of the camera are known, you can find the 3D points through triangulation; conversely, if you do not know the calibration parameters of the camera, you cannot compute the projection relations using normalized image. The projective geometry method allows the projection matrix and 3D points to be estimated using only corresponding points in different views.[128] The problem of computing a 3D reconstruction from two views of the camera can be formulated as finding the projection matrix M by

$$\begin{pmatrix} x \\ y \\ 1 \end{pmatrix} = \begin{pmatrix} m_{11} & m_{12} & m_{13} & m_{14} \\ m_{21} & m_{22} & m_{23} & m_{24} \\ m_{31} & m_{32} & m_{33} & m_{34} \end{pmatrix} \begin{pmatrix} X \\ Y \\ Z \\ 1 \end{pmatrix}$$ (9.11)

$$\begin{pmatrix} (x^a m_{31}^a - m_{11}^a) & (x^a m_{32}^a - m_{12}^a) & (x^a m_{33}^a - m_{13}^a) & (x^a m_{34}^a - m_{14}^a) \\ (y^a m_{31}^a - m_{21}^a) & (y^a m_{32}^a - m_{22}^a) & (y^a m_{33}^a - m_{23}^a) & (y^a m_{34}^a - m_{24}^a) \\ (x^b m_{31}^b - m_{11}^b) & (x^b m_{32}^b - m_{12}^b) & (x^b m_{33}^b - m_{13}^b) & (x^b m_{34}^b - m_{14}^b) \\ (y^b m_{31}^b - m_{21}^b) & (y^b m_{32}^b - m_{22}^b) & (y^b m_{33}^b - m_{23}^b) & (y^b m_{34}^b - m_{24}^b) \end{pmatrix} \begin{pmatrix} X \\ Y \\ Z \\ 1 \end{pmatrix} = 0$$ (9.12)

in which (x, y) is a 2D point on the X–Y plane and (X, Y, Z) is a 3D point on the X–Y–Z plane. Both above equations describe the pixel coordinates in the 3D scene when you know the image coordinates in the two views and the 12 coefficients m_i of the projection matrix. Then, obtain two new equations with three additional unknowns in 3D by

$$x = \frac{m_{11}X + m_{12}Y + m_{13}Z + m_{14}}{m_{31}X + m_{32}Y + m_{33}Z + m_{34}}$$

$$y = \frac{m_{21}X + m_{22}Y + m_{23}Z + m_{24}}{m_{31}X + m_{32}Y + m_{33}Z + m_{34}}$$ (9.13)

The projection point of the first view of the camera is at the origin and the image plane is located at the unit distance along the Z axis. The projection point of the second view of the camera is at (X_0, Y_0, Z_0). It has the rotation R and λ is the scale factor in the unknown calibration matrix (K). According to the eq 9.14, the projection matrix is obtained for two different views of image, by

$$
\begin{pmatrix} x^a \\ y^a \\ 1 \end{pmatrix} = \lambda K \begin{pmatrix} 1 & 0 & 0 & 0 \\ 0 & 1 & 0 & 0 \\ 0 & 0 & 1 & 0 \end{pmatrix} \begin{pmatrix} X \\ Y \\ Z \\ 1 \end{pmatrix}
$$

$$
\begin{pmatrix} x^b \\ y^b \\ 1 \end{pmatrix} = \lambda K R \begin{pmatrix} 1 & 0 & 0 & -X_0 \\ 0 & 1 & 0 & -Y_0 \\ 0 & 0 & 1 & -Z_0 \end{pmatrix} \begin{pmatrix} X \\ Y \\ Z \\ 1 \end{pmatrix} \tag{9.14}
$$

For projective reconstruction from uncalibrated cameras, the computation of projective and affine transform between the two views is very important in the process of 3D reconstruction. As such, regardless of camera parameters, it is assumed that the internal calibration parameters of the camera are known. The calibration matrix (K) contains the position of image center, also known as the focal length, f, of the camera, and the principal point (u_0, v_0). It is defined by

$$
K = K_a = K_b = \begin{bmatrix} f & 0 & u_0 \\ 0 & f & v_0 \\ 0 & 0 & 1 \end{bmatrix} \tag{9.15}
$$

It is possible to compute the external parameters of the camera (rotation of the second camera relative to the first), using projected image data by

$$R_x = \begin{pmatrix} 1 & 0 & 0 \\ 0 & \cos(\theta) & \sin(\theta) \\ 0 & -\sin(\theta) & \cos(\theta) \end{pmatrix}$$

$$R_y = \begin{pmatrix} \cos(\theta) & 0 & \sin(\theta) \\ 0 & 1 & 0 \\ -\sin(\theta) & 0 & \cos(\theta) \end{pmatrix}$$

$$R_z = \begin{pmatrix} \cos(\theta) & \sin(\theta) & 0 \\ -\sin(\theta) & \cos(\theta) & 0 \\ 0 & 0 & 1 \end{pmatrix} \tag{9.16}$$

in which θ is the angle of rotation between the second view and the first view. Any rotation in 3D can be decomposed into simple rotations like these. Therefore, a general rotation can be expressed as a 3×3 matrix of

$$R = R_X R_Y R_Z \tag{9.17}$$

In the final step, the texture mapping is applied to the 3D point cloud to provide a 3D surface of the SEM image. The texture mapping of the 3D model is a method for adding surface texture to a computer-generated model in computer graphics. The textures for the 3D objects are generated by using an automatically projection method. The texture coordinates are equivalent to the XYZ coordinates in the 3D spaces. Basically, the texture mapping specifies the connection between the points in a 3D object with the position of the image on the object. Due to the optical limitation of the measuring instrument, the datasets often contain nonmeasured data points (holes or voids) that requires being filled before any further analysis. The virtual surface-based 3D model was converted into a volume-based model by adding 3D points to the surface mesh with the software program.[128]

9.2.3.1 3D SEM RECONSTRUCTION

In this study, it is assumed that images are obtained by perspective projection and the camera is always uncalibrated and its internal parameters are unknown. The ideas in this chapter can be seen as reversing the rules for 3D modeling using 2D images with no information on the cameras being used. Generally, for 3D reconstruction of a single 2D image, at least two views of the image are needed. In this work, we have prepared two different views in five different regions of the nanofibrous mat in different magnification of camera for 3D reconstruction of single 2D image by rotating the object in three positions while the camera is fixed (Fig. 9.20).

According to Table 9.8, the low magnification with longer working distance leads to a better detection of feature points and improves the depth of focus. It is important to improve the depth of focus at low magnification because insufficient focusing of regions far away from the image center may lead the reconstruction procedure to fail at the edges. Improving the depth of focus is particularly important for the microscopes which do not compensate for the beam rotation induced by a change of focus since the function of dynamic focusing cannot be used in such microscopes. In addition, the low magnification with a long working distance has some problem and it may lead to decrease the field distortion parallel to the tilt axis.[129]

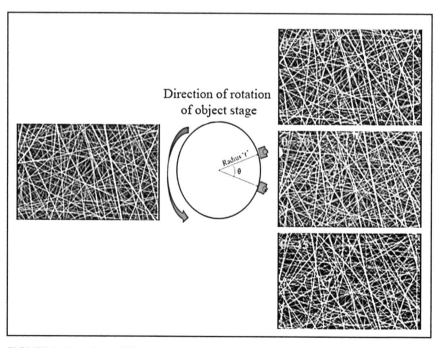

FIGURE 9.20   Three different views of nanofibrous membrane.

TABLE 9.8 Number of Matched Points in Three Different Views of Samples.

Angle of rotation (°)	Number of regions	Number of matched points		
		PAN	PVA	PU
15	1	357	250	654
	2	342	244	697
	3	248	295	670
	4	335	407	670
	5	440	383	616
30	1	68	138	209
	2	100	128	229
	3	63	125	250
	4	88	178	161
	5	117	171	175
45	1	54	119	90
	2	72	106	114
	3	58	94	97
	4	64	109	84
	5	58	118	74

Generally, an increasing angle between of two views decreases the number of matched points. Because, the larger the angle, the bigger the perspective distortions. So, it is getting more difficult for the matching algorithm to identify points correctly, and thus, the probability of the wrong match increases. These differences are then a source of error in depth estimation. Therefore, the angle of rotation should be kept as small as possible in order to overcome the problem of the quick disappearance of the objects in successive views. For this reason, least angle is the appropriate angle for generation of two views of the image.

For obtaining a 3D point cloud, we need to compute the 3D position associated with each match. For this end, the Delaunay triangulation is an excellent way that is computed using the CGAL. It provides our data points as grouped into sets of points that belong to the same global surface, evaluates the projection of each triangle of the Delaunay triangulation in the chosen views, computes the mean of the color variance of the pixels in this triangle, and obtains 3D position of the feature points to modify a generic model using a geometrical deformation.

The texture mapping of the 3D model is a method for adding surface texture to a computer-generated model in computer graphics. The textures for the 3D objects are generated using an automatically projection method. The texture coordinates are equivalent to the XYZ coordinates in the 3D spaces. Basically, the texture mapping specifies the connection between the points in a 3D object with the position of the image on the object.

To the best of our knowledge, the implemented methodology is the first to include photoconsistency testing into a Digital Surf MountainMap software (Mountains Map, Digital Surf, Besancon, France) volumetric reconstruction method. The following subsections describe the steps of the novel algorithm, pointing out the main differences relatively to the standard voxel-based volumetric methods. Once the 3D dataset is acquired from each optical instrument, it is then processed using the Digital Surf MountainMap (v5) software. This software is a common software solution widely used in research and industries for analysis of 3D surface data acquired from various instruments. The data collected by SEM was reconstructed into a 3D image using MountainsMap® Imaging Topography software.

The applied 3D steps on three different nanofibrous mats are illustrated in Figures 9.21–9.23. These figures present 3D point clouds, 3D surface meshes, and the shape models from different perspectives. 2D images were obtained by tilting the sample stage 15° from one to the next in the image sequence in magnification of 2500×.

9.3 VERIFICATION

9.3.1 ROUGHNESS OF NANOFIBROUS MAT AND COMPARISON OF METHODS

The height was measured with a profilometer on the nanofibrous mat samples (Fig. 9.24). The height of nanofibrous mat in different regions of the surface was simulated by relating the grayscale level of the images to the measured height of sample.

By this method, a more appropriate function is obtained to compute height values in all the surface points. Finally, according to eq 9.3, the surface roughness is calculated from the vertical distance between the highest and lowest heights of the profile. The height histograms (Fig. 9.25)

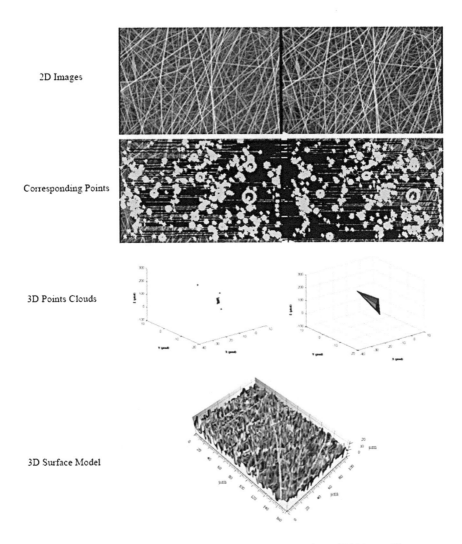

FIGURE 9.21   Visualization of 3D SEM surface reconstruction of PAN nanofibrous mat.

show the statistical distribution of height values and quantify the uniformity on the nanofibrous mats.

For further investigations of the structure, morphology, and surface roughness, the AFM was carried out. 3D surface roughnesses were measured on the nanofibrous mats. The scan size by AFM can be chosen by any

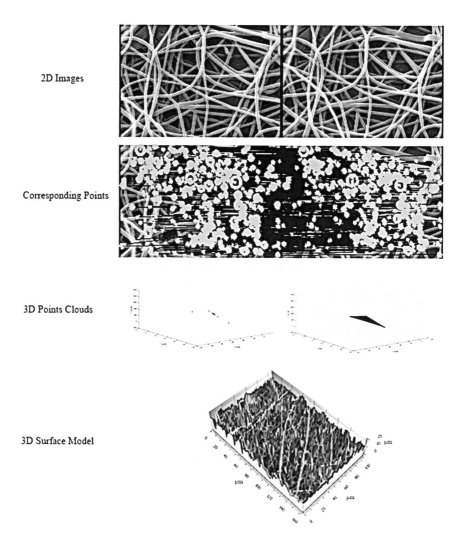

2D Images

Corresponding Points

3D Points Clouds

3D Surface Model

FIGURE 9.22   Visualization of 3D SEM surface reconstruction of PVA nanofibrous mat.

value up to 100 μm. Each measurement contains 256×256 data points. Figure 9.26 shows the AFM images and surface profiles of two different nanofibrous mats on a standard sampling area of $L \times L = 5 \times 5 \ \mu m^2$.

In Figure 9.27, the bar graph represents the average surface roughness for the proposed method and direct measurement by profilometer and AFM. Figure 9.27 illustrates that the electrospun PAN nanofibrous

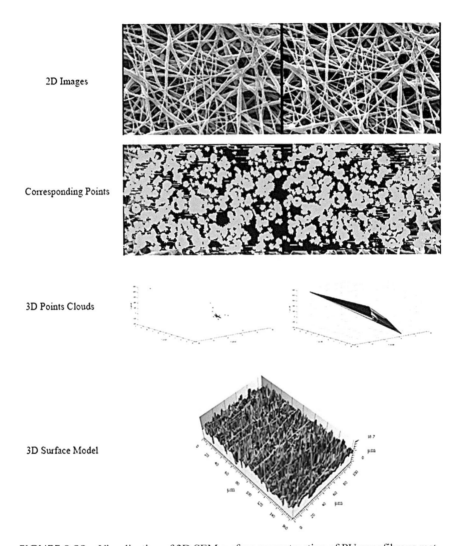

2D Images

Corresponding Points

3D Points Clouds

3D Surface Model

FIGURE 9.23 Visualization of 3D SEM surface reconstruction of PU nanofibrous mat.

membrane with finer fiber diameters has a bigger surface roughness as compared to bigger fiber diameters since the thinner diameters correspond to higher specific surface areas of nanofibrous membrane and rugged surface of nanofibers. Also, comparison of the roughness of nanofibrous mats between direct measurement and proposed method illustrated a satisfactorily close agreement with a low coefficient of expansion.

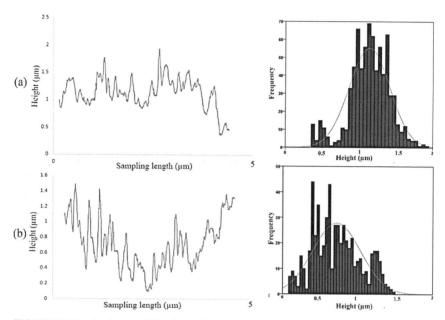

FIGURE 9.24 Surface profile of nanofibrous mats made by profilometer in two different nanofibrous mat with diameters: (a) 187 nm and (b) 213 nm.

9.3.2 THICKNESS OF NANOFIBROUS MAT AND COMPARISON OF METHODS

The cross-sectional view of the nanofibrous mats and the thickness distribution are illustrated in Figure 9.28. The direct measurement in different regions of the electrospun mat is used to verify the result of the proposed image analysis-based method.

Figure 9.29 shows the linear regression with high coefficient of determination ($R^2 = 0.96$) between the average thicknesses of nanofibrous mats obtained by the proposed method and direct measurements. The slope is almost 1 (slope = 1.063), which means that the two methods give the same thickness values.

9.3.3 POROSITY OF NANOFIBROUS MAT AND COMPARISON OF METHODS

Pore structural properties including porosity of different samples can be easily calculated from the 3D reconstructed images. Porosity of three

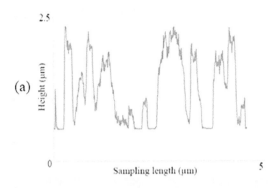

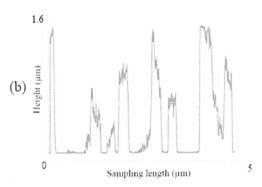

FIGURE 9.25 Surface profile of nanofibrous mats made by proposed method in two different nanofibrous mat with diameters: (a) 187 nm and (b) 213 nm.

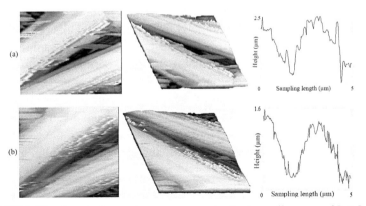

FIGURE 9.26 Atomic force microscopy on a standard sampling area of L × L = 5 × 5 μm² for in two different nanofibrous mat with diameters: (a) 187 nm and (b) 213 nm.

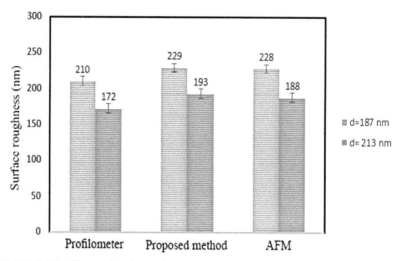

FIGURE 9.27 **(See color insert.)** Average surface roughness (Ra) for the proposed method and direct measurement by profilometer and AFM.

different nanofibrous mats are tabulated in Table 9.9 by using three approaches, that is, 3D model, 2D model, and pycnometer. The results for each sample are based on an average value of five replicate samples. According to Table 9.9, increasing the density resulted in a decrease in porosity, whereas the number of pores, total number of fibers, and indeed total number of crossovers increased with the density raise.

According to Figure 9.30, there is a good agreement between the results obtained by 3D model and pycnometer. Although the results calculated by the model are more useful to predict the porosities in the samples, their theoretical values do not fall far short of the experimental values. This difference between the results obtained from the theory and experiments might be due to underestimation of the model for the pore size due to the limitations caused by the assumptions which have been taken to model the nanofibrous structures as it is stated by the authors. Also, the comparison between 2D and 3D method demonstrated excellence and high accuracy of 3D proposed method.

Figure 9.31 shows the linear regression result between the porosity obtained by the two methods. The result shows significant linear relationship with high coefficient of determination ($R^2 = 0.9993$). It is noted that

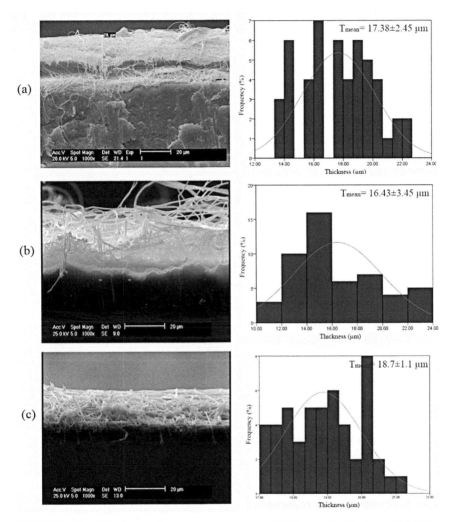

FIGURE 9.28   SEM image shows a cross-section and thickness distribution of (a) PAN, (b) PVA, and (c) PU electrospun nanofibrous mat.

the average porosity obtained from pycnometer are in reasonable agreement with the volumetric porosities obtained from proposed method for the samples. The slope is almost 1 (slope = 0.9982), which means that the two methods gives the same values. It is the reason that the statistically significant number of images in several different regions of nanofibrous mats is used to get its average porosity.

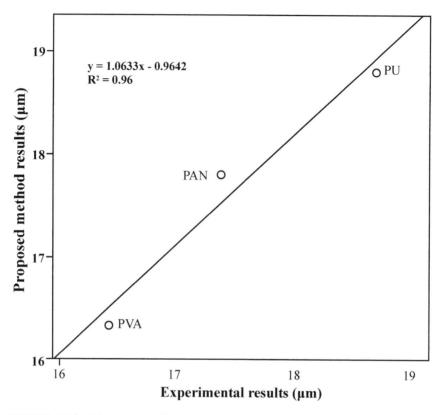

FIGURE 9.29 Linear regression between average thicknesses of nanofibrous mats obtained by the two methods.

TABLE 9.9 Calculated Average Fiber Diameter, Density, and Porosity of Three Different Nanofibrous Mats Using 3D Model, 2D Model, and Pycnometer.

Samples	Diameter (nm)	Thickness (μm)	Density (g/cm³)	Porosity (%)		
				3D model	2D model	Pycnometer
PAN	260	17.38	0.63	43.20	35.20	43.8
PVA	677	16.43	0.33	60.23	47.57	60
PU	593	18.7	0.29	75.89	53.45	76

PAN, polyacrylonitrile; PU, polyurethane; PVA, polyvinyl alcohol.

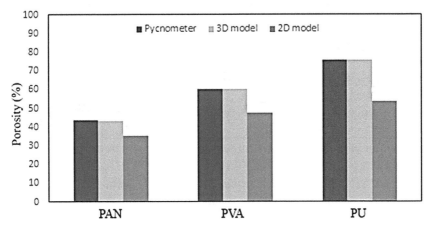

FIGURE 9.30 Comparison of volumetric porosities and pore area fractions of specimens.

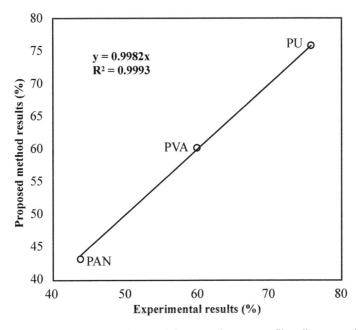

FIGURE 9.31 The linear regression result between the average fiber diameters obtained by the two methods.

9.4   CONCLUSIONS

A novel 3D model based on image analysis technique for estimating the structure parameters of electrospun nanofibrous mat, such as roughness, thickness, and porosity, was developed. The 3D structure was reconstructed from the 2D SEM image of two views of nanofibrous mat, by depth estimation of nanofibrous mat (third dimension) without using any 3D instruments. The method was applied to several samples of nanofibrous membrane in five different regions of mat, with three different angles between of two viewpoints of SEM image. A set of 3D points is computed from two SEM images of a surface, and then a 3D surface is reconstructed from these points using the Delaunay triangulation. A very realistic 3D surface is obtained by applying texture mapping using the Digital Surf MountainMap (v5) software. The resulting points were reconstructed and the errors were evaluated and discussed. The accuracy of the reconstruction obtained depends on the number of viewpoint of images used, the positions of each viewpoint considered, the angle between of two viewpoints of SEM image, and the complexity of the object's shape. Therefore, the high accuracy was obtained in the low magnification and small angle between two viewpoints of SEM image. To evaluate the results, direct measurements were applied on the nanofibrous mat. It was found that the proposed method provides the same value as that of the direct methods. The evidence showed that the proposed method will be useful for extraction of volume information, such as porosity and thickness of nanofibrous structures with high precision and accuracy. It reveals that the developed method can be used as a practical tool to estimate structure parameters of nanofibrous mats.

KEYWORDS

- renewable resource
- polylactic acid
- carbon nanotubes
- nanocomposite
- biomedical applications

REFERENCE

1. Barhate, R. S.; Ramakrishna, S. Nanofibrous Filtering Media: Filtration Problems and Solutions from Tiny Materials. *J. Membr. Sci.* **2007**, *296* (1), 1–8.

2. Cho, D.; Naydich, A.; Frey, M. W.; Joo, Y. L. Further Improvement of Air Filtration Efficiency of Cellulose Filters Coated with Nanofibers via Inclusion of Electrostatically Active Nanoparticles. *Polymer* **2013**, *54* (9), 2364–2372

3. Jena, A.; Gupta, K. Characterization of Pore Structure of Filtration Media. *Fluid/Part. Sep. J.* **2002**, *14* (3), 227–241.

4. Bagherzadeh, R.; Latifi, M.; Najar, S. S.; Kong, L. Three Dimensional Pore Structure Analysis of Nano/Microfibrous Scaffolds Using Confocal Laser Scanning Microscopy. *J. Biomed. Mater. Res. Part A* **2013**, *101* (3), 765–774.

5. Ghasemi-Mobarakeh, L.; Semnani, D.; Morshed, M. A Novel Method for Porosity Measurement of Various Surface Layers of Nanofibers Mat Using Image Analysis for Tissue Engineering Applications. *J. Appl. Polym. Sci.* **2007**, *106*, 2536–2542.

6. Sreedhara, S. S. Rao Tata, N. A Novel Method for Measurement of Porosity in Nanofiber Mat Using Pycnometer in Filtration. *J. Eng. Fibers Fabr.* **2013**, *8* (4),132–137.

7. Borhani, S.; Hosseini S. A.; Etemad, S. G.; Militký, J. Structural Characteristics and Selected Properties of Polyacrylonitrile Nanofiber Mats. *J. Appl. Polym. Sci.* **2008**, *108* (5), 2994–3000.

8. Tiryaki, V. M.; Ayres, V. M.; Khan, A. A.; Ahmed, I.; Shreiber, D. I.; Meiners, S. Nanofibrillar Scaffolds Induce Preferential Activation of Rho GTPases in Cerebral Cortical Astrocytes. *Int. J. Nanomed.* **2012**, *7*, 3891–3905.

9. Ejtehadi, O.; Esfahani, J. A.; Roohi, E. Compressibility and Rarefaction Effects on Entropy and Entropy Generation in Micro/Nano Couette Flow Using DSMC. *J. Phys. Conf. Ser. IOP Publishing* **2012**, *362* (1), 012008 (1–12).

10. Jeyapoovan, T.; Murugan, M.; Surface Roughness Classification Using Image Processing. *Measurement* **2013**, *46* (7), 2065–2072.

11. Azarova, V. V.; Lokhov, U. N.; Malitsky, K. In *Determination of Precise Optical Surface Roughness Parameters Using ARS Data*. International Conference on Applied Optical Metrology (International Society for Optics and Photonics), Italy, 458–465, 1998.

12. Maradudin, A. A. Ed. *Light Scattering and Nanoscale Surface Roughness.* Springer Science & Business Media: USA, 2010.

13. Lehmann, D.; Seidel, F.; Zahn, D. R. Thin Films with High Surface Roughness: Thickness and Dielectric Function Analysis Using Spectroscopic Ellipsometry. *SpringerPlus* **2014**, *3* (1), 1–8.

14. Baltsavias, E. P. A Comparison Between Photogrammetry and Laser Scanning. *ISPRS J. Photogramm. Remote Sens.* **1999**, *54* (2), 83–94.

15. Pavlović, Ž.; Risović, D.; Novaković, D. Comparative Study of Direct and Indirect Image-based Profilometry in Characterization of Surface Roughness. *Surf. Interface Anal.* **2012**, *44* (7), 825–30.

16. Poon, C. Y.; Bhushan, B. Comparison of Surface Roughness Measurements by Stylus Profiler, AFM and Non-contact Optical Profiler. *Wear* **1995**, *190* (1), 76–88.

17. Orji, N. G.; Vorburger, T. V.; Fu, J.; Dixson, R. G.; Nguyen, C. V.; Raja, J. Line Edge Roughness Metrology Using Atomic Force Microscopes. *Meas. Sci. Technol.* **2005,** *16* (11), 2147–2154.

18. Koev, S. T.; Ghodssi, R. Advanced Interferometric Profile Measurements Through Refractive Media. *Rev. Sci. Instrum.* **2008,** *79* (9), 093702.

19. Seitavuopio, P. *The Roughness and Imaging Characterization of Different Pharmaceutical Surfaces.* Faculty of Pharmacy, University of Helsinki: Finland, 2006.

20. Huang, Z. M.; Zhang, Y. Z.; Ramakrishna, S.; Lim, C. T. Electrospinning and Mechanical Characterization of Gelatin Nanofibers. *Polymer* **2004,** *145* (15), 5361–5368.

21. Ziabari, M.; Mottaghitalab, V.; Haghi, A. K. A New Approach for Optimization of Electrospun Nanofiber Formation Process. *Korean J. Chem. Eng.* **2010,** *27,* 340–354.

22. Nasouri, K.; Shoushtari, A. M.; Kaflou, A. Investigation of Polyacrylonitrile Electrospun Nanofibres Morphology as a Function of Polymer Concentration, Viscosity and Berry Number. *Micro. Nano. Lett.* **2012,** *7,* 423–426.

23. Alemdar, A.; Sain, M. Isolation and Characterization of Nanofibers from Agricultural Residues—Wheat Straw and Soy Hulls. *Bioresour. Technol.* **2008,** *99,* (6), 1664–1671.

24. Shrestha, A. Characterization of Porous Membranes via Porometry. MSC Thesis, University of Colorado, 2012.

25. Borkar, N. Characterization of Microporous Membrane Filters Using Scattering Techniques. MSC Thesis, B.S. University of Cincinnati, 2010.

26. Cuperus, F. P.; Smolders, C. A. Characterization of UF Membranes. *Adv. Colloid Interface Sci.* **1991,** *34,* 135–173.

27. Mart´ınez, L.; Florido-D´ıaz, F. J.; Hernández, A.; Prádanos, P. Characterisation of Three Hydrophobic Porous Membranes Used in Membrane Distillation Modelling and Evaluation of Their Water Vapour Permeabilities. *J. Membr. Sci.* **2002,** *203,* 15–27.

28. Bloxson, J. M. Characterization of the Porosity Distribution Within the Clinton Formation. Ashtabula County, Ohio by Geophysical Core and Well Logging, MSC Thesis, Kent State University, 2012.

29. Cao, G. Z.; Meijerink, J.; Brinkman, H. W.; Burggraa, A. J. Permporometry Study on the Size Distribution of Active Pores in Porous Ceramic Membranes. *J. Membr. Sci.* **1993,** *83,* 221–235.

30. Cuperus, F. P.; Bargeman, D.; Smolders, C. A. Permporometry: The Determination of the Size Distribution of Active Pores in UF Membranes. *J. Membr. Sci.* **1992,** *71,* 57–67.

31. Fernando, J. A.; Chuung, D. D. L. Pore Structure and Permeability of an Alumina Fiber Filter Membrane for Hot Gas Filtration. *J. Porous. Mat.* **2002,** *9,* 211–219.

32. Kim, K. J.; Fanen, A. G.; Ben Aimb, R.; Liub, M. G.; Jonsson, G.; I. C.; Tessaro, C.; Broekd, A. P.; Bargeman, D. A Comparative Study of Techniques Used for Porous Membrane Characterization: Pore Characterization. *J. Membr. Sci.* **1994,** *81,* 35–46.

33. Cañas, A.; Ariza, M. J.; Benavente, J. Characterization of Active and Porous Sublayers of a Composite Reverse Osmosis Membrane by Impedance Spectroscopy,

Streaming and Membrane Potentials, Salt Diffusion and X-ray Photoelectron Spectroscopy Measurements. *J. Membr. Sci.* **2001**, *183*, 135–146.

34. Ziabari, M.; Mottaghitalab, V.; Haghi, A. K. Evaluation of Electrospun Nanofiber Pore Structure Parameters. *Korean J. Chem. Eng.* **2008**, *25* (4), 923–932.

35. Miao, J.; Ishikawa, T.; Johnson, B.; Anderson, E. H.; Lai, B.; Hodgson, K. O. High Resolution 3D X-ray Diffraction Microscopy. *Phys. Rev. Lett.* **2002**, *89* (8), 088303.

36. Yan, Lu.; Hui, L.; Xianda, S.; Jianghong, L.; Shuili, Y. Confocal Laser Canning Microscope Analysis of Organic-inorganic Microporous Membranes. *Desalination* **2007**, *217*, 203–211.

37. Rollett, A. D.; Lee, S. B.; Campman, R.; Rohrer, G. S. Three-dimensional Characterization of Microstructure by Electron Back-scatter Diffraction. *Annu. Rev. Mater. Res.* **2007**, *37*, 627–658.

38. Prior, D. J.; Boyle, A. P.; Brenker, F.; Cheadle, M. C.; Day, A.; Lopez, G.; Zetterström, L. The Application of Electron Backscatter Diffraction and Orientation Contrast Imaging in the SEM to Textural Problems in Rocks. *Am. Mineral.* **1999**, *84*, 1741–1759.

39. Davies, P. A.; Randle, V. Combined Application of Electron Backscatter Diffraction and Stereophotogrammetry in Fractography Studies. *J. Microsc.* **2001**, *204* (1), 29–38.

40. Uchic, M. D.; Groeber, M. A.; Dimiduk, D. M.; Simmons, J. P. 3D Microstructural Characterization of Nickel Superalloys via Serial-sectioning Using a Dual Beam FIB-SEM. *Scripta Materialia* **2006**, *55* (1), 23–28.

41. Joosa, J.; Carrarob, T.; Webera, A.; Ivers-Tiffée, E. Reconstruction of Porous Electrodes by FIB/SEM for Detailed Microstructure Modeling. *J. Power Sour.* **2011**, *196*, 7302–7307.

42. Sambaer, W.; Zatloukal, M.; Kimmer, D. The Use of Novel Digital Image Analysis Technique and Rheological Tools to Characterize Nanofiber Nonwovens. *Polym. Test.* **2010**, *29* (1), 82–94.

43. Remondino, F.; El-Hakim, S. Image Based 3D Modelling: A Review. *Photogramm. Rec.* **2006**, *21* (115), 269–291.

44. Kushal, A. M.; Bansal, V.; Banerjee, S. A Simple Method for Interactive 3D Reconstruction and Camera Calibration from a Single View. ICVGIP, 2002.

45. Sturm, P. F.; Maybank, S. J. In *A Method for Interactive 3D Reconstruction of Piecewise Planar Objects from Single Images.* The 10th British Machine Vision Conference (BMVC '99), 265–274, 1999.

46. Criminisi, A.; Reid, I.; Zisserman, A. Single View Metrology. *Int. J. Computer Vision* **2000**, *40* (2), 123–148.

47. Debevec, P. E. Modeling and Rendering Architecture from Photographs. PhD Thesis, University of California at Berkeley, 1996.

48. Grossmann, E.; Ortin, D.; Santos-Victor, J. In *Single and Multi-View Reconstruction of Structured Images.* The 5th Asian Conference on Computer Vision, Taiwan, 23–25, 2002.

49. Liebowitz, D.; Criminisi, A.; Zisserman, A. *Creating Architectural Models from Images.* The Eurographics Association and Blackwell Publishers: UK, 1999; Vol. 18, No. 3, 1–13.

50. Wilczkowiak, M.; Boyer, E.; Sturm, P. In *Camera Calibration and 3D Reconstruction from Single Images Using Parallelepipeds*. 8th International Conference on Computer Vision (ICCV '01), USA, Vol. 1, 142–148, 2001.

51. Pilehrood, M. K.; Heikkil, P. Simulation of Structural Characteristics and Depth Filtration Elements in Interconnected Nanofibrous Membrane Based on Adaptive Image Analysis. *World J. Nano Sci. Eng.* **2013,** *3* (1), 6–16.

52. Sambaer, W.; Zatloukal, M.; Kimmer, D. 3D Modeling of Filtration Process via Polyurethane Nanofiber Based Nonwoven Filters Prepared by Electrospinning Process. *Chem. Eng. Sci.* **2011,** *66* (4), 613–623.

53. Jaganathan, S.; Tafreshi, H. V.; Pourdeyhimi, B. A Realistic Approach for Modeling Permeability of Fibrous Media: 3-D Imaging Coupled with CFD Simulation. *Chem. Eng. Sci.* **2008,** *63* (1), 244–252.

54. Zobel, S.; Maze, B.; Tafreshi, H. V.; Wang, Q.; Pourdeyhimi, B. Simulating Permeability of 3-D Calendered Fibrous Structures. *Chem. Eng. Sci.* **2007,** *62* (22), 6285–6296.

55. Faessel, M.; Delisée, C.; Bos, F.; Castéra, P. 3D Modelling of Random Cellulosic Fibrous Networks Based on X-ray Tomography and Image Analysis. *Compos. Sci. Technol.* **2005,** *65* (13), 1931–1940.

56. Soltani, P.; Johari, M. S.; Zarrebini, M. Effect of 3D Fiber Orientation on Permeability of Realistic Fibrous Porous Networks. *Powder Technol.* **2014,** *254,* 44–56.

57. Hosseini, S. A.; Tafreshi, H. V. Modeling Permeability of 3-D Nanofiber Media in Slip Flow Regime. *Chem. Eng. Sci.* **2010,** *65* (6), 2249–2254.

58. Hosseini, S. A.; Tafreshi, H. V. 3-D Simulation of Particle Filtration in Electrospun Nanofibrous Filters. *Powder Technol.* **2010,** *201* (2), 153–160.

59. Ji, Y.; Ghosh, K.; Shu, X. Z.; Li, B.; Sokolov, J. C.; Prestwich, G. D.; Clark R. A.; Rafailovich, M. H. Electrospun Three-dimensional Hyaluronic Acid Nanofibrous Scaffolds. *Biomaterials* **2006,** *27* (20), 3782–3792.

60. Reingruber, H.; Zankel, A.; Mayrhofer, C.; Poelt, P. Quantitative Characterization of Microfiltration Membranes by 3D Reconstruction. *J. Membr. Sci.* **2011,** *372* (1), 66–74.

61. Ostadi, H.; Rama, P.; Liu, Y.; Chen, R.; Zhang, X. X.; Jiang, K. 3D Reconstruction of a Gas Diffusion Layer and a Microporous Layer. *J. Membr. Sci.* **2010,** *351* (1), 69–74.

62. Pierantonio, F.; Masiero, A.; Bezzo, F.; Beghi, A.; Barolo, M. In *An Improved Multivariate Image Analysis Method for Quality Control of Nanofiber Membranes.* Preprints of the 18th IFAC World Congress Milano, Italy, 2011.

63. Kushal, A. M.; Bansal, V.; Banerjee, S. A Simple Method for Interactive 3D Reconstruction and Camera Calibration from a Single View. ICVGIP, 2002.

64. Sturm, P. F.; Maybank, S. J. A Method for Interactive 3D Reconstruction of Piecewise Planar Objects from Single Images. The 10th British Machine Vision Conference (BMVC '99), UK, 265–274, 1999.

65. Criminisi, A.; Reid, I.; Zisserman, A. Single View Metrology. *Int. J. Computer Vision* **2000,** *40* (2), 123–148.

66. Debevec, P. E. Modeling and Rendering Architecture from Photographs. PhD Thesis, University of California at Berkeley, 1996.

67. Grossmann, E.; Ortin, D.; Santos-Victor, J. In *Single and Multi-View Reconstruction of Structured Images*. The 5th Asian Conference on Computer Vision, Taiwan, 23–25, 2002.

68. Levitz, P. Toolbox for 3D Imaging and Modeling of Porous Media: Relationship with Transport Properties. *Cem. Concr. Res.* **2007,** *37* (3), 351–359.

69. Kushal, A.; Bansal, V.; Banerjee, S. A Simple Method for Interactive 3D Reconstruction, Camera Calibration from a Single View. ICVGIP, 2002.

70. Henrichsen, A. 3D Reconstruction and Camera Calibration from 2D Images. MSc Thesis, University Of Cape Town, 2000.

71. Hartley, R.; Silpa-Anan, C. In *Reconstruction from Two Views Using Approximate Calibration*. Proceedings 5th Asian Conference Computer Vision, Taiwan, 2002.

72. Snavely, N.; Simon, I.; Goesele, M.; Szeliski, R.; Seitz, S. M. In *Scene Reconstruction and Visualization from Community Photo Collections*. Proceedings of the IEEE, USA, Vol. 98, No. 8, 1370–1390, 2010.

73. Zhang, L.; Member, S.; Vázquez, C.; Knorr, S. 3D-TV Content Creation: Automatic 2D-to-3D Video Conversion. *IEEE Trans. Broadcast.* **2011,** *57* (2), 372–383.

74. Andal, F. A.; Taubin, G.; Goldenstein, S. Vanishing Point Detection by Segment Clustering on the Projective Space. Trends and Topics in Computer Vision, Lecture Notes in Computer Science, Vol. 6554, pp 324–33, 2012.

75. Ourselin, S.; Roche, A.; Subsol, G.; Pennec, X.; Ayache, N. Reconstructing a 3D Structure from Serial Histological Sections. *Image Vision Comput.* **2000,** *19,* 25–31.

76. Koura, O. M. Applicability of Image Processing for Evaluation of Surface Roughness. *IOSR J. Eng. (IOSRJEN)* **2015,** *5,* 1–8.

77. Zhongxiang, H.; Lei, Z.; Jiaxu, T.; Xuehong, M.; Xiaojun, S. Evaluation of Three-dimensional Surface Roughness Parameters Based on Digital Image Processing. *Int. J. Adv. Manuf. Technol.* **2009,** *40* (3–4), 342–348.

78. Thornbush, M. J. Measuring Surface Roughness Through the Use of Digital Photography and Image Processing. *Int. J. Geosci.* 2014, 20–24.

79. Lau, K. H. Nanoscale Effects and Applications of Self-organized Nanostructured Thin Films. Doctoral dissertation, Johannes Gutenberg-Universität Mainz, 2008.

80. Banerjee, S.; Yang, R.; Courchene, C. E.; Conners, T. E. Scanning Electron Microscopy Measurements of the Surface Roughness of Paper. *Ind. Eng. Chem. Res.* **2009,** *48* (9), 4322–4325.

81. Koblar, V.; Pecar, M.; Gantar, K.; Tušar, T.; Filipic, B. In *Determining Surface Roughness of Semifinished Products Using Computer Vision and Machine Learning*. Proceedings of the 18th International Multiconference Information Society, Slovenija, IS, 51–54. 2015.

82. Bennett, J. M. Recent Developments in Surface Roughness Characterization. *Meas. Sci. Technol.* **1992,** *3* (12), 1119–1127.

83. Raabe, D.; Sachtleber, M.; Zhao, Z.; Roters, F.; Zaefferer, S. Micromechanical and Macromechanical Effects in Grain Scale Polycrystal Plasticity Experimentation and Simulation. *Acta Materialia* **2001,** *49* (17), 3433–3441.

84. Carrihill, B.; Hummel, R. Experiments with the Intensity Ratio Depth Sensor. *Computer Vision Gr. Image Process.* **1985,** *32* (3), 337–358.

85. Kim, J.; Joy, D. C.; Lee, S. Y. Controlling Resist Thickness and Etch Depth for Fabrication of 3D Structures in Electron-beam Grayscale Lithography. *Microelectron. Eng.* **2007,** *84* (12), 2859–2864.

86. Devi, H. Thresholding: A Pixel-level Image Processing Methodology Preprocessing Technique for an OCR System for the Brahmi Script. *Anc. Asia* **2006,** *1,* 161–165.

87. Mukhopadhyay, S.; Chanda, B. Multiscale Morphological Segmentation of Gray-scale Images. *IEEE Trans. Image Process.* **2003,** *12* (5), 533–549.

88. Moreno, R.; Borga, M.; Smedby, Ö. Estimation of Trabecular Thickness in Grayscale Images Through Granulometric Analysis. *SPIE Med. Imaging (Int. Soc. Optics Photonics),* **2012,** 1–9.

89. Marinello, F.; Bariani, P.; Savio, E.; Horsewell, A.; De Chiffre, L. Critical Factors in SEM 3D Stereo Microscopy. *Meas. Sci. Technol.* **2008,** *19* (6), 065705 (1–12).

90. Hanefeld, P.; Sittner, F.; Ensinger, W.; Greiner, A. Investigation of the Ion Permeability of Poly(p-xylylene) Films. *e-Polymers* **2006,** *6* (1), 341–346.

91. Swanepoel, R. Determination of Surface Roughness and Optical Constants of Inhomogeneous Amorphous Silicon Films. *J. Phys. E: Sci. Instrum.* **1984,** *17,* 896–903.

92. Krizbergs, J.; Kromanis, A. In *Methods for Prediction of the Surface Roughness 3D Parameters According to Technological Parameter*s. 5th International DAAAM Baltic Conference, Tallinn, Estonia, 2006.

93. Tang, X. I.; Xiao, H.; Ding, H.; Liu, J. U. Surface Roughness Measurement Based on Image Processing and Image Recognition. *Comput. Simul. Mod. Sci.* **2009,** 91–96.

94. Li, Z.; Wang, C. Effects of Working Parameters on Electrospinning. In *One-Dimensional Nanostructures*; Springer: Berlin Heidelberg, 2013; pp 15–28.

95. Ganjkhanlou, Y.; Bayandori Moghaddam, A.; Hosseini, S.; Nazari, T.; Gazmeh, A.; Badraghi, J. Application of Image Analysis in the Characterization of Electrospun Nanofibers. *Iran. J. Chem. Chem. Eng. (IJCCE)* **2014,** *33* (2), 37–45.

96. Samah, B.; Charif, D. E. In *Morphological Characterization of SEM Thin Silver Films Using Thresholding*. The First National Conference on Electronics and New Technologies (NCENT'2015), M'Sila, Algeria, 2015.

97. Wong, H. S.; Head, M. K.; Buenfeld, N. R. Pore Segmentation of Cement-based Materials from Backscattered Electron Images. *Cem. Concr. Res. 36* (6), **2006,** 1083–1090.

98. Brent Neal, F.; Russ, J. C. *Measuring Shape*, CRC Press: USA, 2012.

99. Kim, H.; Choi, S.; Sohn, K. In *Real-time Depth Reconstruction from Stereo Sequences*. International Society for Optics and Photonics, Proceedings of SPIE 6016, 60160E, USA, (1–12), 2005.

100. Fahmy, A. A. Stereo Vision Based Depth Estimation Algorithm in Uncalibrated Rectification. *Int. J. Video Image Process Netw Secur.* **2013,** *13* (2), 1–8.

101. Mühlmann, K.; Maier, D.; Hesser, J.; Männer, R. Calculating Dense Disparity Maps from Color Stereo Images, an Efficient Implementation. *Int. J. Computer Vision (IJCV).* **2002,** *47* (1), 79–88.

102. Garg, P.; Yerva, S.; Kutty, K. In *Relative Depth Estimation Using a Rotating Camera System*. Proceedings of the International Conference on Image Processing, Computer Vision, and Pattern Recognition (IPCV), USA, 1–6, 2013.

103. Saxena, A.; Chung, S. H.; Ng, A. Y. 3-d Depth Reconstruction from a Single Still Image. *Int. J. Computer Vision* **2008**, *76* (1), 53–69.

104. Wei, J.; Sugimoto, S.; Okutomi, M. In *Panoramic 3D Reconstruction Using Rotating Camera with Planar Mirrors.* The 6th Workshop on Omnidirectional Vision, Camera Networks and Non-classical Cameras, 1–8, 2005.

105. Lai, S. H.; Fu, C. W.; Chang, S. A Generalized Depth Estimation Algorithm with a Single Image. *IEEE Trans. Pattern Anal. Mach. Intell.* **1992**, *14* (4), 405–411.

106. Sun, C.; Berman, M.; Coward, D.; Osborne, B. Thickness Measurement and Crease Detection of Wheat Grains Using Stereo Vision. *Pattern. Recogn. Lett.* **2007**, *28* (12), 1501–1508.

107. Rahman, W.; Chen, F. K.; Yeoh, J.; Patel, P.; Tufail, A.; Da Cruz, L. Repeatability of Manual Subfoveal Choroidal Thickness Measurements in Healthy Subjects Using the Technique of Enhanced Depth Imaging Optical Coherence Tomography. *Invest. Ophthalmol. Vis. Sci. (IOVS)* **2011**, 52 (5), 2267–2271.

108. Shin, E. H.; Cho, K. S.; Seo, M. H.; Kim, H. Determination of Electrospun Fiber Diameter Distributions Using Image Analysis Processing. *Macromol. Res.* **2008**, *16* (4), 314–319.

109. Pan, B.; Qian, K.; Xie, H.; Asundi, A. Two-dimensional Digital Image Correlation for In-plane Displacement and Strain Measurement: A Review. *Meas. Sci. Technol.* **2009**, *20* (6), 1–17.

110. Alhwarin, F.; Wang, C.; Ristic-Durrant, D.; Gräser, A. In *Improved SIFT-Features Matching for Object Recognition.* BCS International Academic Conference, UK, 178–190, 2008.

111. Swapna, A.; Geetha Devi, A. A Comparison and Matching Point Extraction of SIFT and ISIFT. *Int. J. Comput. Tech. Appl.* **2013**, *4* (6), 1026–1033.

112. Ambrosch, K.; Kubinger, W.; Humenberger, M.; Steininger, A. Flexible Hardware-based Stereo Matching. *EURASIP J. Embed. Sys.* **2008**, *2008*, 1–12.

113. Zhu, F. Y.; Wang, Q. Q.; Zhang, X. S.; Hu, W.; Zhao, X.; Zhang, H. X. 3D Nanostructure Reconstruction Based on the SEM Imaging Principle, and Applications. *Nanotechnology* **2014**, *25* (18), 185705 (1–10).

114. Gravel, P.; Verhaeghe, J.; Reader, A. J. 3D PET Image Reconstruction Including Both Motion Correction and Registration Directly into an MR or Stereotaxic Spatial Atlas. *Phys. Med. Biol.* **2013**, *58* (1), 105–126.

115. Paaver, U.; Heinämäki, J.; Kassamakov, I.; Hæggström, E.; Ylitalo, T.; Nolvi, A.; Veski, P. Nanometer Depth Resolution in 3D Topographic Analysis of Drug-loaded Nanofibrous Mats Without Sample Preparation. *Int. J. Pharm.* **2014**, *462* (1), 29–37.

116. Wiederkehr, T.; Klusemann, B.; Gies, D.; Müller, H.; Svendsen, B. An Image Morphing Method for 3D Reconstruction and FE-analysis of Pore Networks in Thermal Spray Coatings. *Computat. Mater. Sci.* **2010**, *47*, 881–889.

117. Delerue, J. F.; Perrie, E.; Yu, Z. Y.; Velde, B. New Algorithms in 3D Image Analysis and Their Application to the Measurement of a Spatialized Pore Size Distribution in Soils. *Phys. Chem. Earth (A)* **1999**, *24* (7), 639–644.

118. Al-Raoush, R. I.; Willson, C. S. Extraction of Physically Realistic Pore Network Properties from Three-dimensional Synchrotron X-ray Microtomography Images of Unconsolidated Porous Media Systems. *J. Hydrol.* **2005**, *300*, 44–64.

119. Liang, Z. R.; Fernandes, C. P.; Magnani, F. S.; Philippi, P. C. A Reconstruction Technique for Three-dimensional Porous Media Using Image Analysis and Fourier Transforms. *J. Petroleum Sci. Eng.* **1998,** *21,* 273–283.

120. Diógenes, A. N.; dos Santos, L. O. E.; Fernandes, C. P.; Moreira, A. C.; Apolloni, C. R. Porous Media Microstructure Reconstruction Using Pixel-Based and Object-based Simulated Annealing—Comparison with Other Reconstruction Methods. *Engenharia Térmica (Therm. Eng.)* **2009,** *8* (2), 35–41.

121. Faessel, M.; Delisee, C.; Bos, F.; Castera, P. 3D Modelling of Random Cellulosic Fibrous Networks Based on X-ray Tomography and Image Analysis. *Compos. Sci. Technol.* **2005,** *65,* 1931–1940.

122. Mao, L.; Shi, P.; Tu, H.; An, L.; Ju, Y.; Hao, N. Porosity Analysis Based on CT Images of Coal Under Uniaxial Loading. *Adv. Comput. Tomogr.* **2012,** *1,* 5–10.

123. Robin, V.; Sardini, P.; Mazurier, A.; Regnault, O.; Descostes, M. Effective Porosity Measurements of Poorly Consolidated Materials Using Non-destructive Methods. *Eng. Geol.* **2016,** *205,* 24–29.

124. Tafti, A. P.; Kirkpatrick, A. B.; Alavi, Z.; Owen, H. A.; Yu, Z. Recent Advances in 3D SEM Surface Reconstruction. *Micron* 78, 54–66, 2015.

125. Labatut, P.; Pons, J. P.; Keriven, R. In *Efficient Multi-view Reconstruction of Large-Scale Scenes Using Interest Points, Delaunay Triangulation and Graph Cuts.* Computer Vision ICCV 2007, 11th International Conference on IEEE, Switzerland, 1–8, 2007.

126. Oblonsek, C.; Guid, N. A Fast Surface-based Procedure for Object Reconstruction from 3D Scattered Points. *Comput. Vis. Image. Underst.* **1998,** *69,* 185–195.

127. Fua, P.; Sander, P. In *Reconstructing Surfaces from Unstructured 3d Points.* Second European Conference on Computer Vision (ECCV'90), Japan, 1992.

128. Lee, W. S.; Park, J. K.; Kim, J. H.; Kim, H. Y.; Kim, W. C.; Yu, C. H. New Approach to Accuracy Verification of 3D Surface Models: An Analysis of Point Cloud Coordinates. *J. Prosthodont. Res.* **2016,** *60,* 98–105.

129. Pouchou, J. L.; Boivin, D.; Beauchêne, P.; G.; Le Besnerais, F. Vignon, 3d Reconstruction of Rough Surfaces by SEM Stereo Imaging. *Microchim. Acta.* **2002,** *139,* 135–44.

INDEX

T - #0842 - 101024 - C280 - 229/152/12 - PB - 9781774633953 - Gloss Lamination